Scientific Publishing Ltd.™

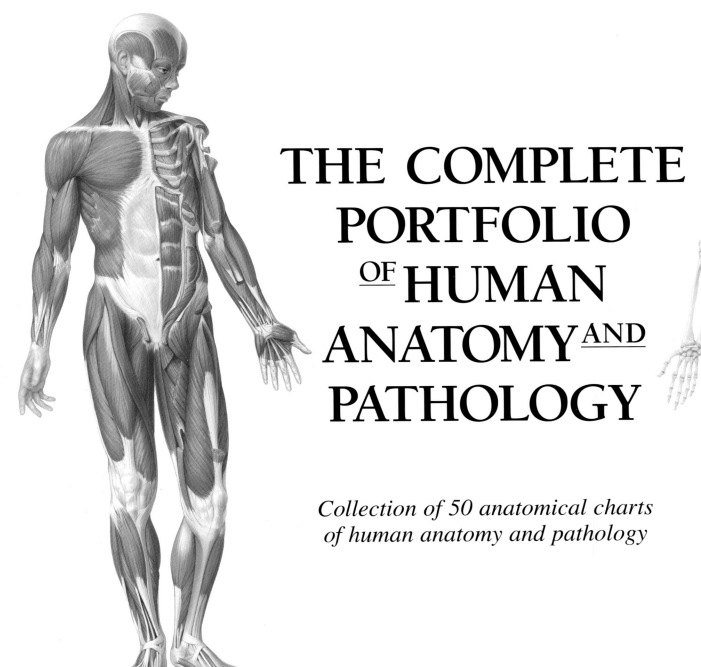

THE COMPLETE PORTFOLIO OF HUMAN ANATOMY AND PATHOLOGY

*Collection of 50 anatomical charts
of human anatomy and pathology*

Scientific Publishing Ltd.
www.scientificpublishing.com

RedShelf

FH27HANJS8BCSUYH

www.RedShelf.com

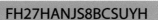

To redeem access code and view eBook follow these steps below:	Required Software	Required Hardware
1. Go to: www.redshelf.com		
2. Click 'Sign In' icon in the upper right corner or 'Register' if it is your first time visiting the site.	Safari 5 or greater Chrome 10 or greater Firefox 3 or greater Internet Explorer 9 or greater	512 MB of RAM 1 GHz Dual Core or 500 MHz Quad Core 50 MB of Hard Drive Space (offline only)
3. Upon registering, sign in and click the 'Library' icon.		
4. Enter the access code within the "Redemption Code" field and select 'Redeem Code'.		

Compatable Devices

The cloud-reader is accessible on any internet-enabled device that supports the above browsers including: Laptops/Desktops, Tablets & Smartphones.

Offline Mode is compatible on: Laptops/Desktops, Tablets & Smartphones.

Published in the United States by
Scientific Publishing Ltd.
167 Joey Drive Elk Grove Village, IL 60007

Individual chart titles are available at www.scientificpublishing.com

©2014 Scientific Publishing Ltd. Elk Grove Village, IL

All rights reserved under International and Pan-American Copyright Conventions. No part of this book may be reproduced or transmitted in any form or by any means, electronic or mechanical, including photocopy, recording or by any information storage and retrieval system, without permission in writing from the publisher.

ISBN-13: 978-1-935612-34-6

Item# CHAP-50DC

Printed in Hong Kong

INTRODUCTION

Since our very origin we have been on an odyssey of self-discovery…an odyssey to solve the mystery of our own body. Prehistoric skulls recently uncovered in what is now Europe indicate that Neolithic man had actually performed successful surgical procedures. Later, the Chinese, in 2600 B.C., conceptualized the heart's role and postulated the circulation of blood. And while sometimes wrong in their interpretations, the ancient Greeks and Romans nevertheless laid the foundation for modern concepts in human anatomy and physiology. From Hammurabi to Hippocrates to Galen, early physicians grappled with questions of structure and pathology as they attempted to unlock our innermost secrets.

For the most part progress has been rather slow and plodding. Then in the latter half of the 20th century the most incredible advancements in evidence-based medicine were made. Since World War II there has been a virtual explosion in our knowledge of molecular biology, physiology, medicine, pharmaceuticals, surgery and pathology. Now as never before we are able to battle, and at times conquer, the threats facing us. And the essence of that explosion is captured in this state of the art 50-plate portfolio.

In this handsome portfolio you will find an inventory of human anatomy and pathology of rare quality. The accuracy of each plate is unsurpassed and the artwork is beautifully designed and rendered. Just take a look at plate 33 and you will see the intricacies of the brain or turn to plate 13 for a detailed depiction of asthma. Page after page overflows with some of the best and most detailed illustrations ever presented in a single volume. Here is more than a mere glimpse into our structure. Here is a work crafted to reveal, through easy-to-understand explanations and illustrations, the very richness of the human body and the various assaults that it must endure.

Ken Hyde, Ph.D.
Professor of Anatomy

LIST OF PLATES

The Skeletal System

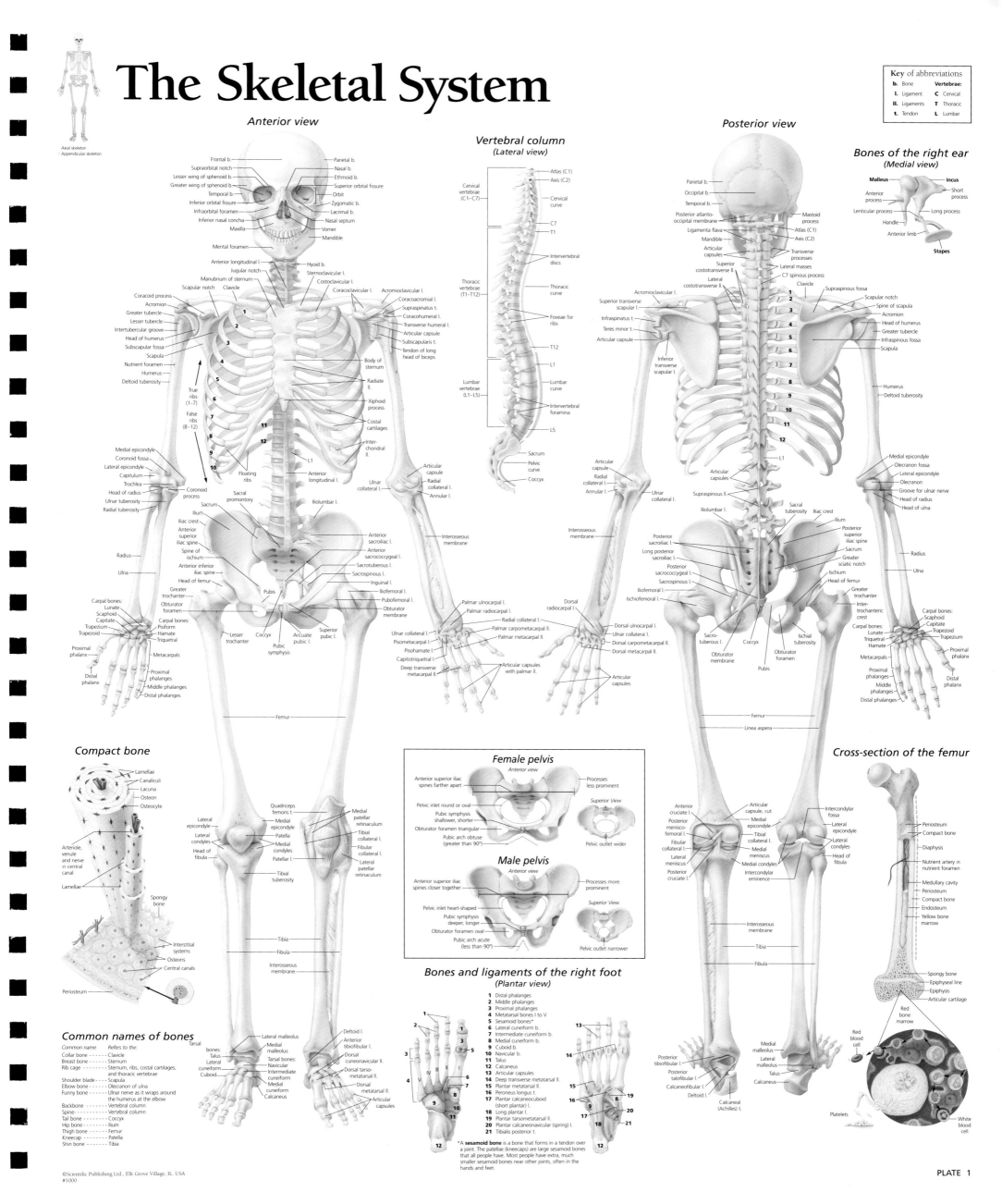

Anterior view

Posterior view

Vertebral column
(Lateral view)

Bones of the right ear
(Medial view)

Key of abbreviations
b. Bone   Vertebrae:
l. Ligament   C Cervical
ll. Ligaments   T Thoracic
t. Tendon   L Lumbar

Compact bone

Female pelvis
Anterior view

Male pelvis
Anterior view

Cross-section of the femur

Common names of bones

Common name   Refers to the:
Collar bone ----- Clavicle
Breast bone ----- Sternum
Rib cage -------- Sternum, ribs, costal cartilages, and thoracic vertebrae
Shoulder blade --- Scapula
Elbow bone ------ Olecranon of ulna
Funny bone ------ Ulnar nerve as it wraps around the humerus at the elbow
Backbone -------- Vertebral column
Spine ----------- Vertebral column
Tail bone -------- Coccyx
Hip bone -------- Ilium
Thigh bone ------ Femur
Kneecap --------- Patella
Shin bone ------- Tibia

Bones and ligaments of the right foot
(Plantar view)

1 Distal phalanges
2 Middle phalanges
3 Proximal phalanges
4 Metatarsal bones I to V
5 Sesamoid bones*
6 Lateral cuneiform b.
7 Intermediate cuneiform b.
8 Medial cuneiform b.
9 Cuboid b.
10 Navicular b.
11 Talus
12 Calcaneus
13 Articular capsules
14 Deep transverse metatarsal ll.
15 Plantar metatarsal ll.
16 Peroneus longus t.
17 Plantar calcaneocuboid (short plantar) l.
18 Long plantar l.
19 Plantar tarsometatarsal ll.
20 Plantar calcaneonavicular (spring) l.
21 Tibialis posterior t.

*A **sesamoid bone** is a bone that forms in a tendon over a joint. The patellae (kneecaps) are large sesamoid bones that all people have. Most people have extra, much smaller sesamoid bones near other joints, often in the hands and feet.

©Scientific Publishing Ltd., Elk Grove Village, IL, USA
#1000

PLATE 1

The Muscular System

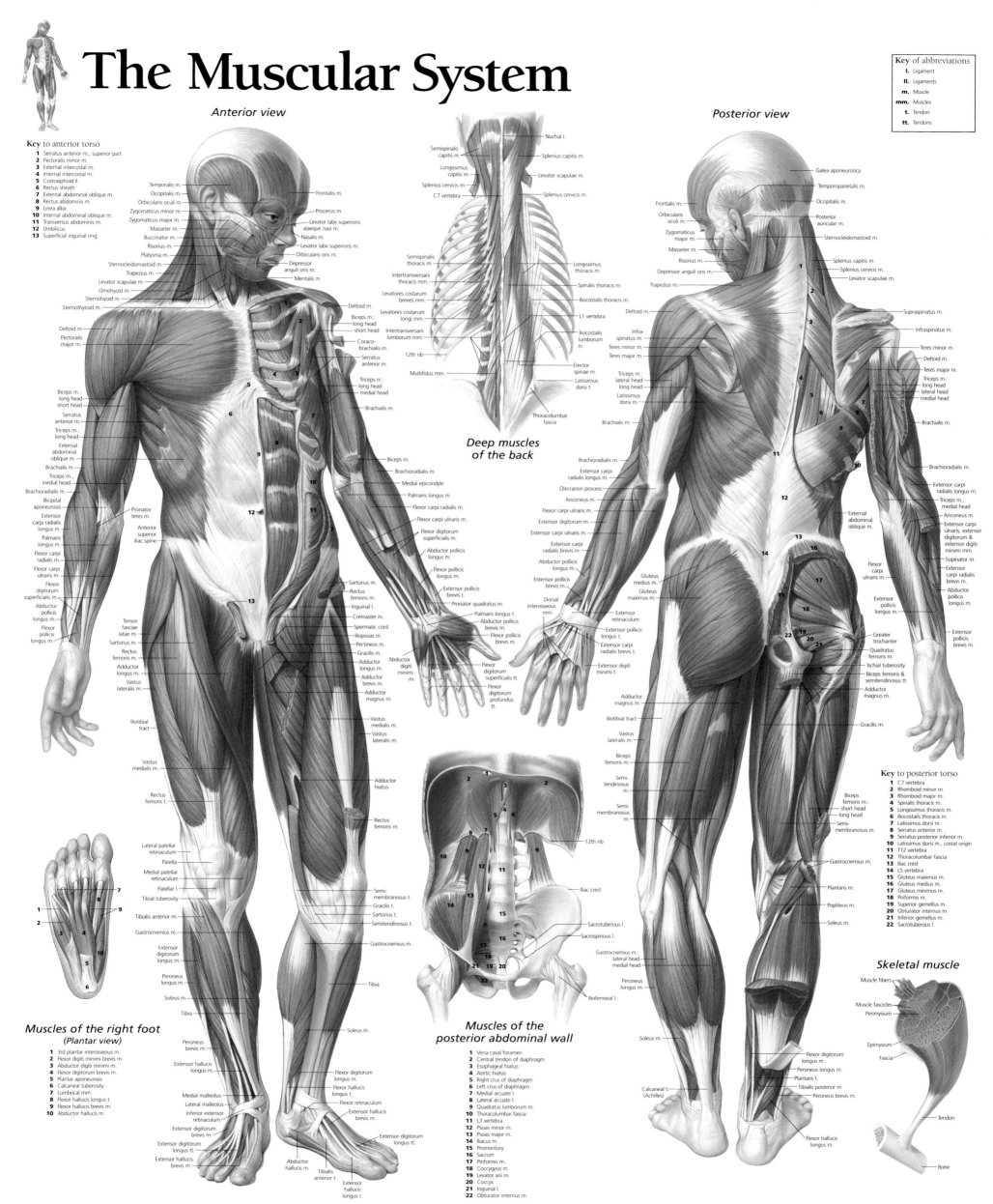

Anterior view

Posterior view

Key of abbreviations
l. Ligament
ll. Ligaments
m. Muscle
mm. Muscles
t. Tendon
tt. Tendons

Key to anterior torso
1 Serratus anterior m., superior part
2 Pectoralis minor m.
3 External intercostal m.
4 Internal intercostal m.
5 Costoxiphoid ll.
6 Rectus sheath
7 External abdominal oblique m.
8 Rectus abdominis m.
9 Linea alba
10 Internal abdominal oblique m.
11 Transversus abdominis m.
12 Umbilicus
13 Superficial inguinal ring

Deep muscles of the back

Muscles of the right foot
(Plantar view)
1 3rd plantar interosseous m.
2 Flexor digiti minimi brevis m.
3 Abductor digiti minimi m.
4 Flexor digitorum brevis m.
5 Plantar aponeurosis
6 Calcaneal tuberosity
7 Lumbrical mm.
8 Flexor hallucis longus t.
9 Flexor hallucis brevis m.
10 Abductor hallucis m.

Muscles of the posterior abdominal wall
1 Vena caval foramen
2 Central tendon of diaphragm
3 Esophageal hiatus
4 Aortic hiatus
5 Right crus of diaphragm
6 Left crus of diaphragm
7 Medial arcuate l.
8 Lateral arcuate l.
9 Quadratus lumborum m.
10 Thoracolumbar fascia
11 L3 vertebra
12 Psoas minor m.
13 Psoas major m.
14 Iliacus m.
15 Promontory
16 Sacrum
17 Piriformis m.
18 Coccygeus m.
19 Levator ani m.
20 Coccyx
21 Inguinal l.
22 Obturator internus m.

Key to posterior torso
1 C7 vertebra
2 Rhomboid minor m.
3 Rhomboid major m.
4 Spinalis thoracis m.
5 Longissimus thoracis m.
6 Iliocostalis thoracis m.
7 Latissimus dorsi m.
8 Serratus anterior m.
9 Serratus posterior inferior m.
10 Latissimus dorsi m., costal origin
11 T12 vertebra
12 Thoracolumbar fascia
13 Iliac crest
14 L5 vertebra
15 Gluteus maximus m.
16 Gluteus medius m.
17 Gluteus minimus m.
18 Piriformis m.
19 Superior gemellus m.
20 Obturator internus m.
21 Inferior gemellus m.
22 Sacrotuberous l.

Skeletal muscle

©Scientific Publishing Ltd., Elk Grove Village, IL. USA
#1100

PLATE 2

The Female Muscular System

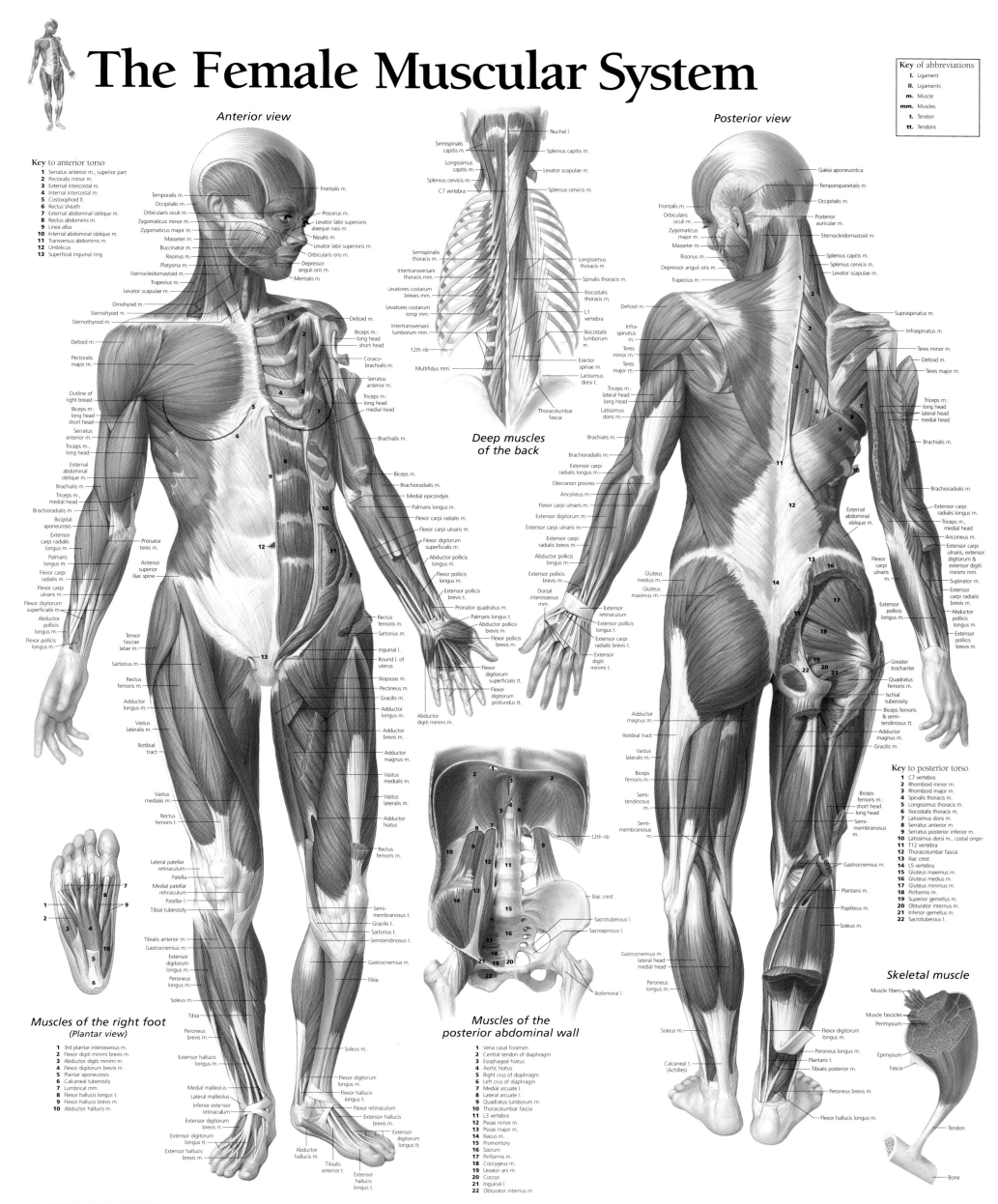

Anterior view

Posterior view

Key of abbreviations
l. Ligament
ll. Ligaments
m. Muscle
mm. Muscles
t. Tendon
tt. Tendons

Deep muscles of the back

Muscles of the right foot
(Plantar view)

Muscles of the posterior abdominal wall

Skeletal muscle

Key to anterior torso
1 Serratus anterior m., superior part
2 Pectoralis minor m.
3 External intercostal m.
4 Internal intercostal m.
5 Costoxiphoid ll.
6 Rectus sheath
7 External abdominal oblique m.
8 Rectus abdominis m.
9 Linea alba
10 Internal abdominal oblique m.
11 Transversus abdominis m.
12 Umbilicus
13 Superficial inguinal ring

Key to posterior torso
1 C7 vertebra
2 Rhomboid minor m.
3 Rhomboid major m.
4 Spinalis thoracis m.
5 Longissimus thoracis m.
6 Iliocostalis thoracis m.
7 Latissimus dorsi m.
8 Serratus anterior m.
9 Serratus posterior inferior m.
10 Latissimus dorsi m., costal origin
11 T12 vertebra
12 Thoracolumbar fascia
13 Iliac crest
14 L5 vertebra
15 Gluteus maximus m.
16 Gluteus medius m.
17 Gluteus minimus m.
18 Piriformis m.
19 Superior gemellus m.
20 Obturator internus m.
21 Inferior gemellus m.
22 Sacrotuberous l.

Muscles of the right foot (Plantar view)
1 3rd plantar interosseous m.
2 Flexor digiti minimi brevis m.
3 Abductor digiti minimi m.
4 Flexor digitorum brevis m.
5 Plantar aponeurosis
6 Calcaneal tuberosity
7 Lumbrical m.
8 Flexor hallucis longus t.
9 Flexor hallucis brevis m.
10 Abductor hallucis m.

Muscles of the posterior abdominal wall
1 Vena caval foramen
2 Central tendon of diaphragm
3 Esophageal hiatus
4 Aortic hiatus
5 Right crus of diaphragm
6 Left crus of diaphragm
7 Medial arcuate l.
8 Lateral arcuate l.
9 Quadratus lumborum m.
10 Thoracolumbar fascia
11 L3 vertebra
12 Psoas minor m.
13 Psoas major m.
14 Iliacus m.
15 Promontory
16 Sacrum
17 Piriformis m.
18 Coccygeus m.
19 Levator ani m.
20 Coccyx
21 Inguinal l.
22 Obturator internus m.

©Scientific Publishing Ltd., Elk Grove Village, IL USA
#1101

PLATE 3

The Hand & Wrist

Anatomy of the hand and wrist

The **hand** and **wrist** together form a complex structure composed of 27 small bones connected by joints, muscles and tendons. The framework of the hand is shaped by five large **metacarpal** bones that articulate with the fingers and wrist. The fingers each contain three smaller jointed bones called **phalanges**; the thumb has two phalanges. The hand's wide range of fine movements is made possible by numerous small joints, long and short muscles, and **extensor** and **flexor** tendons that allow the fingers and thumb to straighten, bend and flex. The highly flexible thumb joint (**carpometacarpal**) is positioned at a 90-degree angle to the finger joints, giving the hand the unique ability to grasp, pinch and manipulate objects. **Tendon sheaths** surrounding the tendons contain **synovial fluid** for smooth movement of the hand and wrist.

The **wrist** connects the hand to the **ulna** and **radius** of the arm. It consists of eight **carpal** bones with multiple joints that allow flexion, extension and rotary movements. A distinctive feature of the wrist is the **carpal tunnel channel** through which the tendons from the hand pass to reach the forearm.

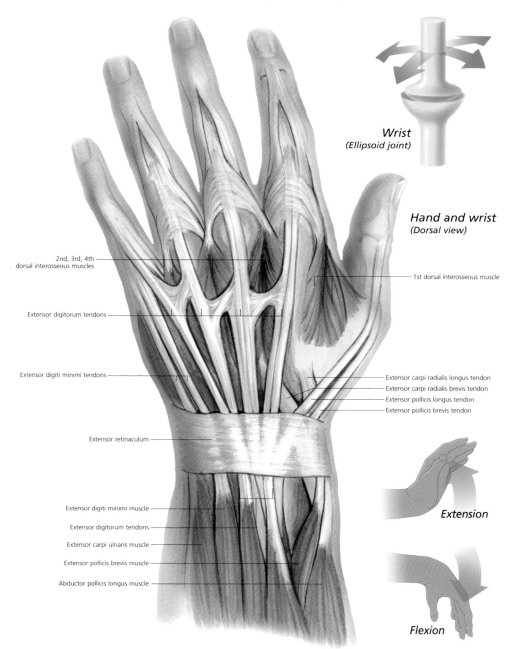

Wrist
(Ellipsoid joint)

Hand and wrist
(Dorsal view)

2nd, 3rd, 4th dorsal interosseous muscles

1st dorsal interosseous muscle

Extensor digitorum tendons

Extensor digiti minimi tendons

Extensor carpi radialis longus tendon
Extensor carpi radialis brevis tendon
Extensor pollicis longus tendon
Extensor pollicis brevis tendon

Extensor retinaculum

Extension

Extensor digiti minimi muscle
Extensor digitorum tendons
Extensor carpi ulnaris muscle
Extensor pollicis brevis muscle
Abductor pollicis longus muscle

Flexion

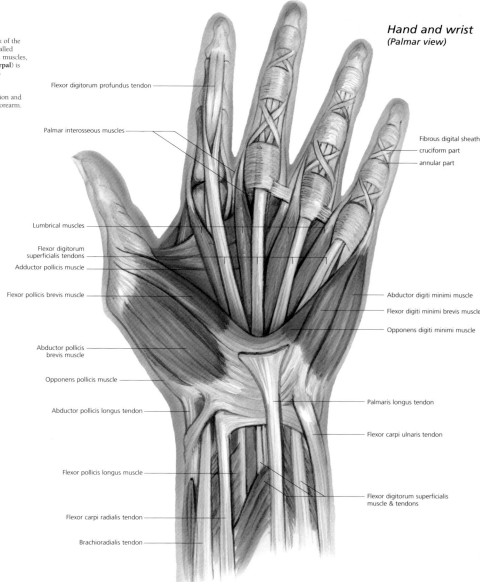

Hand and wrist
(Palmar view)

Flexor digitorum profundus tendon

Palmar interosseous muscles

Fibrous digital sheath:
cruciform part
annular part

Lumbrical muscles
Flexor digitorum superficialis tendons
Adductor pollicis muscle
Flexor pollicis brevis muscle

Abductor digiti minimi muscle
Flexor digiti minimi brevis muscle
Opponens digiti minimi muscle

Abductor pollicis brevis muscle

Opponens pollicis muscle
Abductor pollicis longus tendon

Palmaris longus tendon
Flexor carpi ulnaris tendon

Flexor pollicis longus muscle

Flexor digitorum superficialis muscle & tendons

Flexor carpi radialis tendon

Brachioradialis tendon

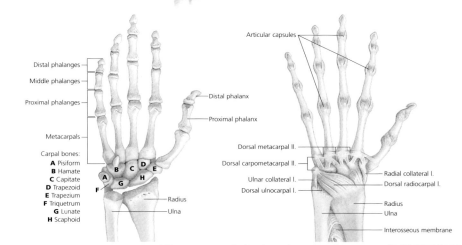

Bones and ligaments of the hand and wrist
(Palmar view)

Articular capsules

Distal phalanges
Middle phalanges
Proximal phalanges

Distal phalanx
Proximal phalanx

Metacarpals

Carpal bones:
A Pisiform
B Hamate
C Capitate
D Trapezoid
E Trapezium
F Triquetrum
G Lunate
H Scaphoid

Radius
Ulna

Dorsal metacarpal II.
Dorsal carpometacarpal II.
Ulnar collateral I.
Dorsal ulnocarpal I.

Radial collateral I.
Dorsal radiocarpal I.
Radius
Ulna
Interosseous membrane

Key of abbreviations
I. Ligament   **II.** Ligaments

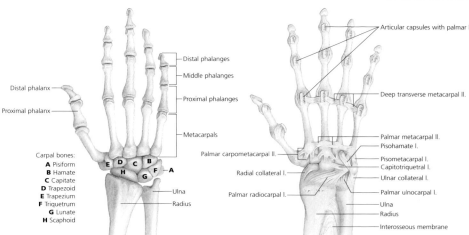

Bones and ligaments of the hand and wrist
(Dorsal view)

Articular capsules with palmar II.

Distal phalanges
Middle phalanges
Proximal phalanges

Distal phalanx
Proximal phalanx

Metacarpals

Carpal bones:
A Pisiform
B Hamate
C Capitate
D Trapezoid
E Trapezium
F Triquetrum
G Lunate
H Scaphoid

Ulna
Radius

Deep transverse metacarpal II.
Palmar metacarpal II.
Palmar carpometacarpal II.
Radial collateral I.
Palmar radiocarpal I.

Pisohamate I.
Pisometacarpal I.
Capitotriquetral I.
Ulnar collateral I.
Palmar ulnocarpal I.
Ulna
Radius
Interosseous membrane

Carpal tunnel syndrome

Carpal tunnel syndrome is an inflammatory condition affecting the median nerve and flexor tendons that pass through the **carpal tunnel**. This archway is formed by the carpal bones and sits beneath the broad **transverse ligament** extending across the palm. Inflammation and thickening of the tendons within the tunnel and the ligament above it may be caused by many factors (see right). Swelling in the tunnel compresses the median nerve, restricting blood flow and oxygen supply and causing tingling, numbness, weakness and pain in the **wrist**, **hand**, **fingers** and **thumb**. Carpal tunnel syndrome can usually be reversed if treated early. Chronic inflammation may lead to permanent nerve damage.

Potential causes of carpal tunnel syndrome include:
- Repetitive stress injuries (overuse syndrome)
- Underlying medical conditions including arthritis, diabetes and obesity
- Pregnancy
- Bone dislocations and fractures

Affected area

Sensory distribution of median nerve

Types of impairment
- Paresthesia
  (abnormal sensation)
- Hypoesthesia
  (diminished sensation)
- Anesthesia
  (partial or total loss of sensation)

Transverse carpal ligament
(flexor retinaculum)
Palmar carpal ligament
Median nerve

(Palmar view)

Colles' fracture

Radius
Fracture

Scaphoid fracture

Scaphoid
Fracture

Fractures

A fracture is a crack or break in a bone. There are multiple bones in the hand and wrist vulnerable to fracture, including the **carpal bones** located at the base of the hand and the **radius** and **ulna**, where they connect to the wrist.

The two most common types of fracture are Colles' and scaphoid fractures. A **Colles' fracture** is a complete transverse break in the end of the radius, and often occurs when the hand is flexed to stop a fall. It is a common injury in older people. A fall on the palm of the hand can also cause a **scaphoid fracture**, a break in the scaphoid carpal bone that articulates with the radius. This injury may initially be confused with a bad sprain and tends to heal slowly due to limited blood supply.

©Scientific Publishing Ltd., Elk Grove Village, IL USA
#1005

PLATE 4

The Foot & Ankle

Ankle –
(Hinge joint)

The foot and ankle

The foot and ankle form a complex structure that includes 33 joints, 26 bones and more than 100 muscles, tendons and ligaments. The feet and ankles work together to provide the body with support, balance and mobility.

Structurally, the **foot** is divided into three sections. The **forefoot** plays a major role in weight-bearing and balance and contains the long bones of the foot (**metatarsals**) and the toes (**phalanges**), which are connected at the ball of the foot by five metatarsal phalangeal joints. The **midfoot** contains five interlocking **tarsal bones** (cuboid, navicular, and 3 cuneiform) that form the arch of the foot. It is connected to the forefoot and hindfoot by muscles and the **plantar fascia** ligament, an important structure that stabilizes the foot and helps maintain the arch. Two additional tarsal bones make up the **hindfoot**. The **calcaneus** (heel) is the largest, strongest bone in the foot and the site of attachment for the powerful **Achilles tendon**. The **talus** (astragalus) sits above the calcaneus and between the lower ends of the leg bones (**tibia** and **fibula**) to form the ankle joint. The talus is involved in multiple planes of movement and is responsible for transferring weight and pressure from the leg to the foot.

The ankle joint itself is a uniaxial, hinge-type joint capable of both upwards (**dorsiflexion**) and downwards (**plantarflexion**) motion. Limited rotation, abduction and adduction is also possible. The ankle joint is protected by a fibrous capsule and supported on each side by strong collateral ligaments.

Key of abbreviations
b. Bone   **l.** Ligament
ll. Ligaments   **t.** Tendon

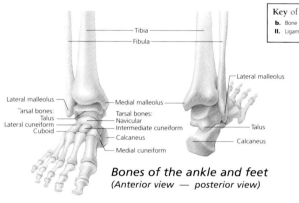

Bones of the ankle and feet
(Anterior view — posterior view)

Foot bones –
(Glide joints)

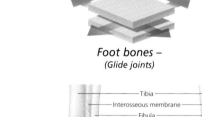

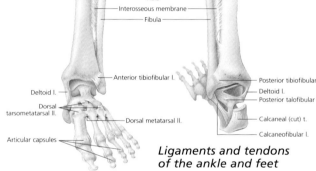

Ligaments and tendons of the ankle and feet
(Anterior view — posterior view)

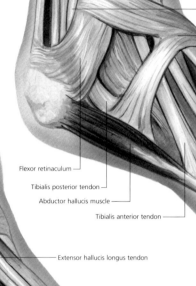

Foot and ankle
(Anterior view)

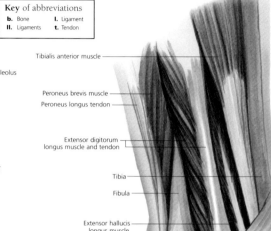

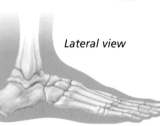

Lateral view

Medial view

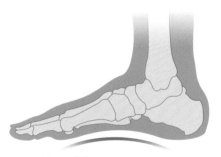

Normal foot –
The longitudinal arch in the foot helps support the body as we stand or walk

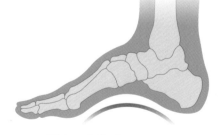

High arch foot –
Also called *pes cavus*, can be caused by muscle imbalances in the foot

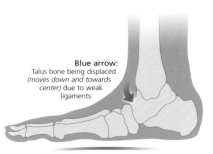

Blue arrow:
Talus bone being displaced *(moves down and towards center)* due to weak ligaments

Flatfoot –
Also known as *pes planus*, occurs when the longitudinal arch is gradually lost or never develops

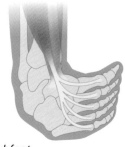

Clubfoot –
Is a congenital deformity (present at birth) from incorrectly formed bones and joints

Bones and ligaments of the left foot
(Plantar view)

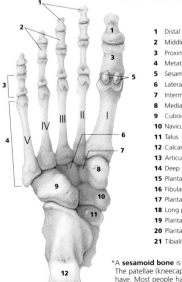

1 Distal phalanges
2 Middle phalanges
3 Proximal phalanges
4 Metatarsal bones I to V
5 Sesamoid bones*
6 Lateral cuneiform b.
7 Intermediate cuneiform b.
8 Medial cuneiform b.
9 Cuboid b.
10 Navicular b.
11 Talus
12 Calcaneus
13 Articular capsules
14 Deep transverse metatarsal ll.
15 Plantar metatarsal ll.
16 Fibularis longus t.
17 Plantar calcaneocuboid (short plantar) l.
18 Long plantar l.
19 Plantar tarsometatarsal ll.
20 Plantar calcaneonavicular (spring) l.
21 Tibialis posterior t.

*A **sesamoid bone** is a bone that forms in a tendon over a joint. The patellae (kneecaps) are large sesamoid bones that all people have. Most people have extra, much smaller sesamoid bones near other joints, often in the hands and feet.

©Scientific Publishing Ltd., Elk Grove Village, IL. USA
#1004

PLATE 5

The Shoulder & Elbow

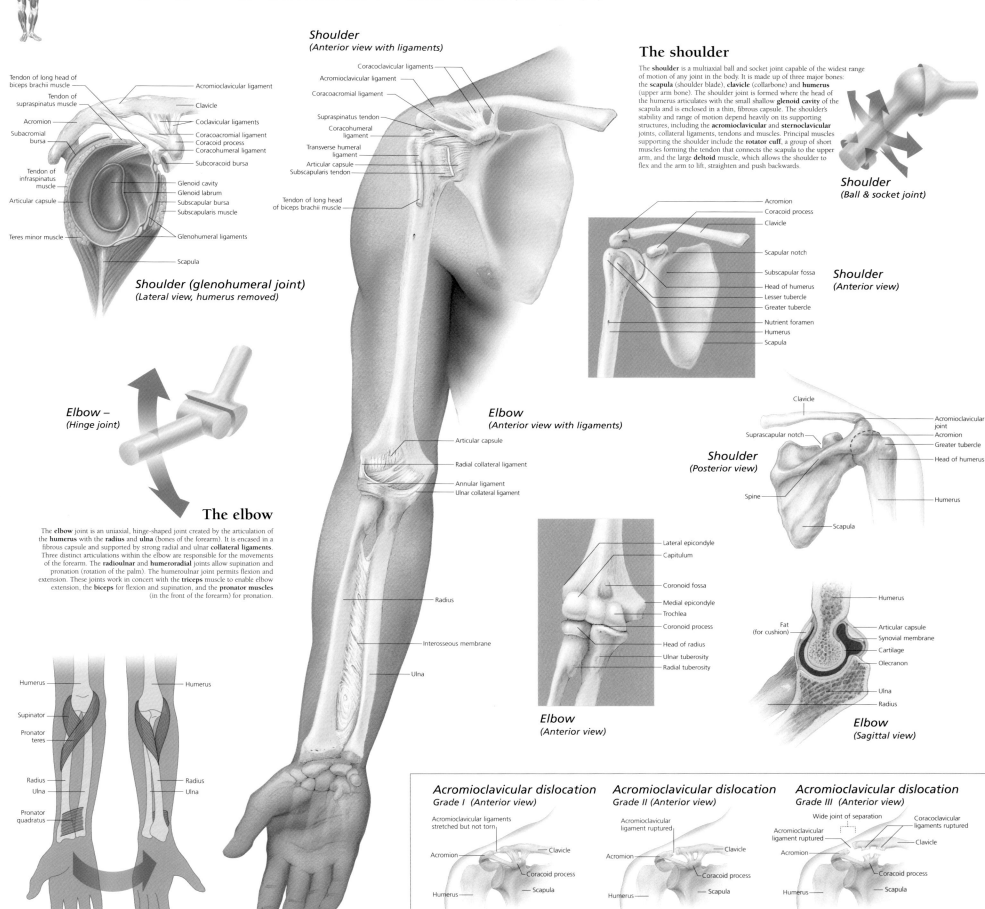

Shoulder
(Anterior view with ligaments)

Coracoclavicular ligaments
Acromioclavicular ligament
Coracoacromial ligament
Supraspinatus tendon
Coracohumeral ligament
Transverse humeral ligament
Articular capsule
Subscapularis tendon
Tendon of long head of biceps brachii muscle

Shoulder (glenohumeral joint)
(Lateral view, humerus removed)

Tendon of long head of biceps brachii muscle
Tendon of supraspinatus muscle
Acromion
Subacromial bursa
Tendon of infraspinatus muscle
Articular capsule
Teres minor muscle
Acromioclavicular ligament
Clavicle
Coclavicular ligaments
Coracoacromial ligament
Coracoid process
Coracohumeral ligament
Subcoracoid bursa
Glenoid cavity
Glenoid labrum
Subscapular bursa
Subscapularis muscle
Glenohumeral ligaments
Scapula

The shoulder

The **shoulder** is a multiaxial ball and socket joint capable of the widest range of motion of any joint in the body. It is made up of three major bones: the **scapula** (shoulder blade), **clavicle** (collarbone) and **humerus** (upper arm bone). The shoulder joint is formed where the head of the humerus articulates with the small shallow **glenoid cavity** of the scapula and is enclosed in a thin, fibrous capsule. The shoulder's stability and range of motion depend heavily on its supporting structures, including the **acromioclavicular** and **sternoclavicular** joints, collateral ligaments, tendons and muscles. Principal muscles supporting the shoulder include the **rotator cuff**, a group of short muscles forming the tendon that connects the scapula to the upper arm, and the large **deltoid** muscle, which allows the shoulder to flex and the arm to lift, straighten and push backwards.

Shoulder
(Ball & socket joint)

Shoulder
(Anterior view)

Acromion
Coracoid process
Clavicle
Scapular notch
Subscapular fossa
Head of humerus
Lesser tubercle
Greater tubercle
Nutrient foramen
Humerus
Scapula

Shoulder
(Posterior view)

Clavicle
Suprascapular notch
Spine
Scapula
Acromioclavicular joint
Acromion
Greater tubercle
Head of humerus
Humerus

Elbow –
(Hinge joint)

The elbow

The **elbow** joint is an uniaxial, hinge-shaped joint created by the articulation of the **humerus** with the **radius** and **ulna** (bones of the forearm). It is encased in a fibrous capsule and supported by strong radial and ulnar **collateral ligaments**. Three distinct articulations within the elbow are responsible for the movements of the forearm. The **radioulnar** and **humeroradial** joints allow supination and pronation (rotation of the palm). The humeroulnar joint permits flexion and extension. These joints work in concert with the **triceps** muscle to enable elbow extension, the **biceps** for flexion and supination, and the **pronator muscles** (in the front of the forearm) for pronation.

Elbow
(Anterior view with ligaments)

Articular capsule
Radial collateral ligament
Annular ligament
Ulnar collateral ligament

Radius
Interosseous membrane
Ulna

Elbow
(Anterior view)

Lateral epicondyle
Capitulum
Coronoid fossa
Medial epicondyle
Trochlea
Coronoid process
Head of radius
Ulnar tuberosity
Radial tuberosity

Elbow
(Sagittal view)

Humerus
Fat (for cushion)
Articular capsule
Synovial membrane
Cartilage
Olecranon
Ulna
Radius

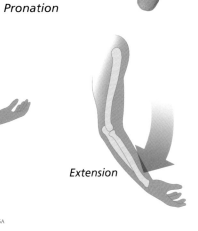

Humerus
Supinator
Pronator teres
Radius
Ulna
Pronator quadratus
Humerus
Radius
Ulna

Supination   **Pronation**

Flexion   **Extension**

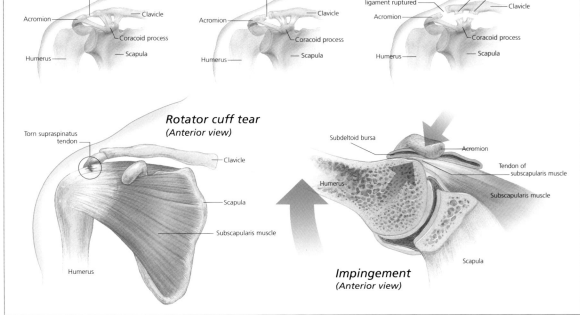

Acromioclavicular dislocation
Grade I *(Anterior view)*

Acromioclavicular ligaments stretched but not torn
Acromion
Humerus
Clavicle
Coracoid process
Scapula

Acromioclavicular dislocation
Grade II *(Anterior view)*

Acromioclavicular ligament ruptured
Acromion
Humerus
Clavicle
Coracoid process
Scapula

Acromioclavicular dislocation
Grade III *(Anterior view)*

Wide joint of separation
Acromioclavicular ligament ruptured
Acromion
Humerus
Coracoclavicular ligaments ruptured
Clavicle
Coracoid process
Scapula

Rotator cuff tear
(Anterior view)

Torn supraspinatus tendon
Clavicle
Scapula
Subscapularis muscle
Humerus

Impingement
(Anterior view)

Subdeltoid bursa
Acromion
Tendon of subscapularis muscle
Subscapularis muscle
Humerus
Scapula

©Scientific Publishing Ltd., Elk Grove Village, IL. USA
#1003

PLATE 6

The Hip & Knee

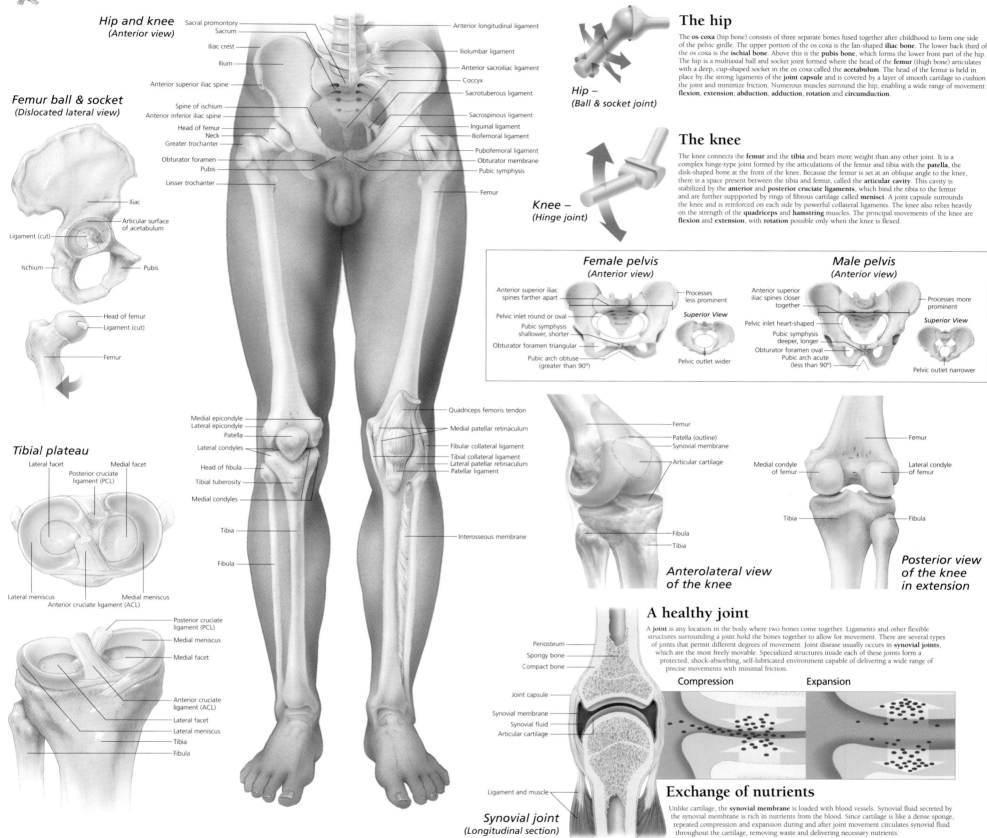

Hip and knee
(Anterior view)

- Sacral promontory
- Sacrum
- Iliac crest
- Ilium
- Anterior superior iliac spine
- Spine of ischium
- Anterior inferior iliac spine
- Head of femur
- Neck
- Greater trochanter
- Obturator foramen
- Pubis
- Lesser trochanter

- Anterior longitudinal ligament
- Iliolumbar ligament
- Anterior sacroiliac ligament
- Coccyx
- Sacrotuberous ligament
- Sacrospinous ligament
- Inguinal ligament
- Iliofemoral ligament
- Pubofemoral ligament
- Obturator membrane
- Pubic symphysis
- Femur

Femur ball & socket
(Dislocated lateral view)

- Iliac
- Articular surface of acetabulum
- Ligament (cut)
- Ischium
- Pubis

- Head of femur
- Ligament (cut)
- Femur

The hip

The **os coxa** (hip bone) consists of three separate bones fused together after childhood to form one side of the pelvic girdle. The upper portion of the os coxa is the fan-shaped **iliac bone**. The lower back third of the os coxa is the **ischial bone**. Above this is the **pubis bone**, which forms the lower front part of the hip. The hip is a multiaxial ball and socket joint formed where the head of the **femur** (thigh bone) articulates with a deep, cup-shaped socket in the os coxa called the **acetabulum**. The head of the femur is held in place by the strong ligaments of the **joint capsule** and is covered by a layer of smooth cartilage to cushion the joint and minimize friction. Numerous muscles surround the hip, enabling a wide range of movement: **flexion, extension, abduction, adduction, rotation** and **circumduction**.

Hip –
(Ball & socket joint)

The knee

The knee connects the **femur** and the **tibia** and bears more weight than any other joint. It is a complex hinge-type joint formed by the articulations of the femur and tibia with the **patella**, the disk-shaped bone at the front of the knee. Because the femur is set at an oblique angle to the knee, there is a space present between the tibia and femur, called the **articular cavity**. This cavity is stabilized by the **anterior** and **posterior cruciate ligaments**, which bind the tibia to the femur and are further supported by rings of fibrous cartilage called **menisci**. A joint capsule surrounds the knee and is reinforced on each side by powerful collateral ligaments. The knee also relies heavily on the strength of the **quadriceps** and **hamstring** muscles. The principal movements of the knee are **flexion** and **extension**, with **rotation** possible only when the knee is flexed.

Knee –
(Hinge joint)

Female pelvis
(Anterior view)

- Anterior superior iliac spines farther apart
- Pelvic inlet round or oval
- Pubic symphysis shallower, shorter
- Obturator foramen triangular
- Pubic arch obtuse (greater than 90°)
- Processes less prominent
- *Superior View*
- Pelvic outlet wider

Male pelvis
(Anterior view)

- Anterior superior iliac spines closer together
- Pelvic inlet heart-shaped
- Pubic symphysis deeper, longer
- Obturator foramen oval
- Pubic arch acute (less than 90°)
- Processes more prominent
- *Superior View*
- Pelvic outlet narrower

Tibial plateau

- Lateral facet
- Medial facet
- Posterior cruciate ligament (PCL)
- Lateral meniscus
- Anterior cruciate ligament (ACL)
- Medial meniscus

- Posterior cruciate ligament (PCL)
- Medial meniscus
- Medial facet
- Anterior cruciate ligament (ACL)
- Lateral facet
- Lateral meniscus
- Tibia
- Fibula

Anterior view labels:
- Medial epicondyle
- Lateral epicondyle
- Patella
- Lateral condyles
- Head of fibula
- Tibial tuberosity
- Medial condyles
- Tibia
- Fibula

- Quadriceps femoris tendon
- Medial patellar retinaculum
- Fibular collateral ligament
- Tibial collateral ligament
- Lateral patellar retinaculum
- Patellar ligament
- Interosseous membrane

Anterolateral view of the knee

- Femur
- Patella (outline)
- Synovial membrane
- Articular cartilage
- Fibula
- Tibia

Posterior view of the knee in extension

- Femur
- Medial condyle of femur
- Lateral condyle of femur
- Tibia
- Fibula

A healthy joint

A **joint** is any location in the body where two bones come together. Ligaments and other flexible structures surrounding a joint hold the bones together to allow for movement. There are several types of joints that permit different degrees of movement. Joint disease usually occurs in **synovial joints**, which are the most freely movable. Specialized structures inside each of these joints form a protected, shock-absorbing, self-lubricated environment capable of delivering a wide range of precise movements with minimal friction.

Compression   **Expansion**

- Periosteum
- Spongy bone
- Compact bone
- Joint capsule
- Synovial membrane
- Synovial fluid
- Articular cartilage
- Ligament and muscle

Synovial joint
(Longitudinal section)

Exchange of nutrients

Unlike cartilage, the **synovial membrane** is loaded with blood vessels. Synovial fluid secreted by the synovial membrane is rich in nutrients from the blood. Since cartilage is like a dense sponge, repeated compression and expansion during and after joint movement circulates synovial fluid throughout the cartilage, removing waste and delivering necessary nutrients.

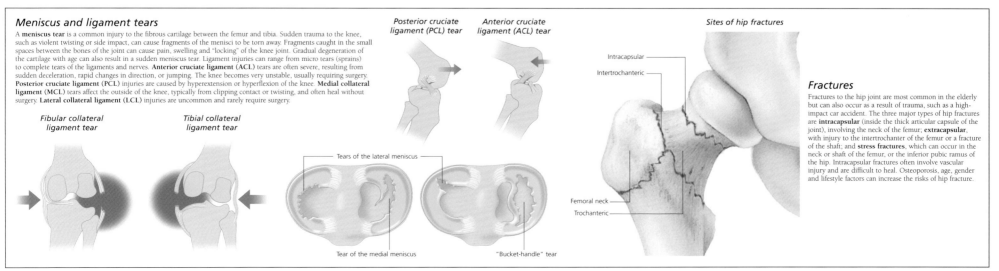

Meniscus and ligament tears

A **meniscus tear** is a common injury to the fibrous cartilage between the femur and tibia. Sudden trauma to the knee, such as violent twisting or side impact, can cause fragments of the menisci to be torn away. Fragments caught in the small spaces between the bones of the joint can cause pain, swelling and "locking" of the knee joint. Gradual degeneration of the cartilage with age can also result in a sudden meniscus tear. Ligament injuries can range from micro tears (sprains) to complete tears of the ligaments and nerves. **Anterior cruciate ligament (ACL)** tears are often severe, resulting from sudden deceleration, rapid changes in direction, or jumping. The knee becomes very unstable, usually requiring surgery. **Posterior cruciate ligament (PCL)** injuries are caused by hyperextension or hyperflexion of the knee. **Medial collateral ligament (MCL)** tears affect the outside of the knee, typically from clipping contact or twisting, and often heal without surgery. **Lateral collateral ligament (LCL)** injuries are uncommon and rarely require surgery.

Fibular collateral ligament tear

Tibial collateral ligament tear

Posterior cruciate ligament (PCL) tear

Anterior cruciate ligament (ACL) tear

- Tears of the lateral meniscus
- Tear of the medial meniscus
- "Bucket-handle" tear

Sites of hip fractures

- Intracapsular
- Intertrochanteric
- Femoral neck
- Trochanteric

Fractures

Fractures to the hip joint are most common in the elderly but can also occur as a result of trauma, such as a high-impact car accident. The three major types of hip fractures are **intracapsular** (inside the thick articular capsule of the joint), involving the neck of the femur; **extracapsular**, with injury to the intertrochanter of the femur or a fracture of the shaft; and **stress fractures**, which can occur in the neck or shaft of the femur, or the inferior pubic ramus of the hip. Intracapsular fractures often involve vascular injury and are difficult to heal. Osteoporosis, age, gender and lifestyle factors can increase the risks of hip fracture.

©Scientific Publishing Ltd., Elk Grove Village, IL. USA
#1002

PLATE 7

Understanding the Spine

- Axial skeleton
- Appendicular skeleton

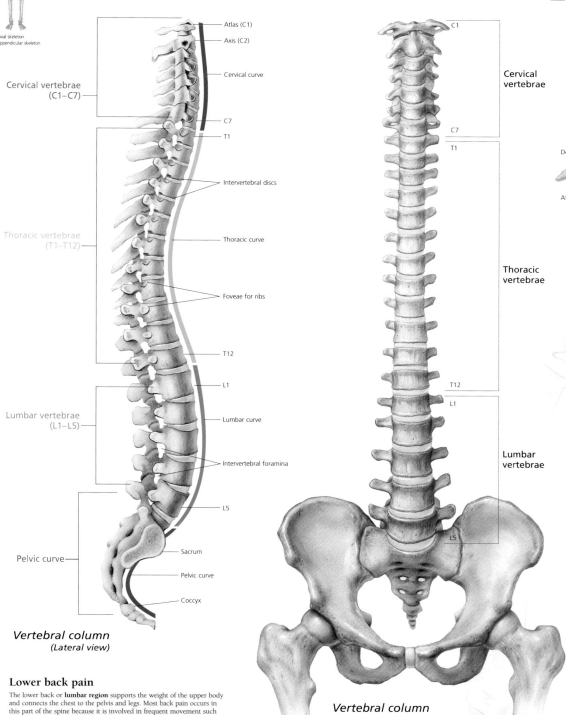

Cervical vertebrae (C1–C7)

- Atlas (C1)
- Axis (C2)
- Cervical curve
- C7
- T1
- Intervertebral discs

Thoracic vertebrae (T1–T12)

- Thoracic curve
- Foveae for ribs
- T12
- L1

Lumbar vertebrae (L1–L5)

- Lumbar curve
- Intervertebral foramina
- L5

Pelvic curve

- Sacrum
- Pelvic curve
- Coccyx

Vertebral column
(Lateral view)

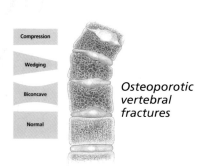

- C1
- Cervical vertebrae
- C7
- T1
- Thoracic vertebrae
- T12
- L1
- Lumbar vertebrae
- L5

Vertebral column
(Anterior view)

The spine

The spine is a column of 26 bones extending from the base of the skull to the pelvis. It is made up of 24 **vertebrae**: 7 **cervical** (neck), 12 **thoracic** (upper back), and 5 **lumbar** (lower back). The **sacrum** lies at the base of the lumbar region and is fused to the **coccyx** (tail bone). The spine provides support for the head, shoulders and chest and protects the **spinal cord**, a long, fragile structure composed of nerves that transmit signals between the brain and body, enabling movement and sensation. As the spinal cord passes through the **vertebral foramen** at the center of the vertebrae, pairs of **spinal nerves** (roots) enter and emerge between the vertebral spaces, connecting with nerves throughout the body. The spinal column is cushioned and protected by **intervertebral discs** made of a spongy outer ring of cartilage and a jelly-like fluid center. The spine is held in place by **ligaments** and **tendons** that attach to bony processes at the back (posterior) of the vertebrae and connect to the muscles of the back.

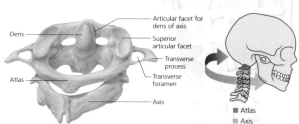

- Dens
- Atlas
- Articular facet for dens of axis
- Superior articular facet
- Transverse process
- Transverse foramen
- Axis
- Atlas
- Axis

Atlas and axis

The 2 uppermost vertebrae of the cervical spine play an important role in the motion and flexibility of the head. The atlas (C1 vertebra), keeps the head supported and enables up and down (nodding) motion while preventing twisting. Articulation with the axis (C2 vertebra) allows side-to-side or rotating movement of the head. Powerful muscles connected to the spinous process of the axis control the position of both the head and neck.

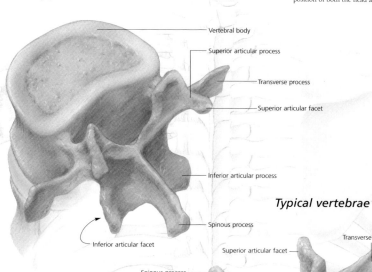

- Vertebral body
- Superior articular process
- Transverse process
- Superior articular facet
- Inferior articular process
- Spinous process
- Inferior articular facet

Typical vertebrae

- Transverse process
- Superior articular facet
- Spinous process
- Superior articular process
- Pedicle
- Inferior articular facet
- Vertebral foramen
- Vertebral body

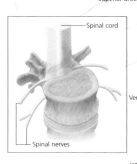

- Spinal cord
- Spinal nerves

- Intervertebral disc

Lower back pain

The lower back or **lumbar region** supports the weight of the upper body and connects the chest to the pelvis and legs. Most back pain occurs in this part of the spine because it is involved in frequent movement such as bending, twisting, turning, standing, walking and lifting. Back pain is often classified as either **acute** (lasting from a few days to 3 months) or **chronic** (lasting 3 months or longer) and may range in severity from a dull ache to shooting or stabbing pain that limits motion, flexibility and/or the ability to stand upright. Because discs weaken with age, low back pain is usually age-related, occurring most frequently in adults through the mid-sixties.

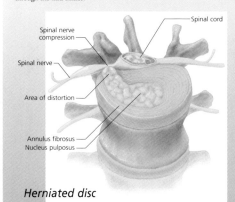

- Spinal cord
- Spinal nerve compression
- Spinal nerve
- Area of distortion
- Annulus fibrosus
- Nucleus pulposus

Herniated disc

Causes of lower back pain

Most acute lower back pain occurs as the result of stress on the muscles and ligaments that support the spine. A sedentary lifestyle and obesity also increase the risks of back injury and pain. Chronic back pain may be caused by underlying illness such as arthritis or depression. Treatment of back pain varies depending on cause and severity but may include heat or ice application, limited exercise, over-the-counter or prescription medications, or in extreme cases, surgery.

Typical causes of acute back pain include:

- Ruptured (herniated) disc
- Muscle strains or sprains
- Degenerative disc disease
- Spinal stenosis
- Sciatica
- Sacroiliitis

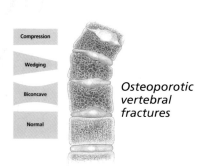

Lumbar vertebrae L1-4 with osteoarthritis
(Lateral view)

Changes in joint shape cause painful compression of surrounding nerves

What is osteoarthritis?

Osteoarthritis (OA), also called **degenerative joint disease** or **osteoarthrosis**, has existed for centuries, occurring in many animals as well as humans. Currently, over 20 million people in the U.S. have OA. It is characterized by a gradual loss of cartilage and overgrowth of bone, often within only one or a few joints. Unlike other forms of arthritis, OA does not spread to other parts of the body. OA can occur in almost any movable joint but most commonly affects the spine, hips, knees, hands or feet.

Risk factors for OA

- Age (risk increases with age, affecting almost everyone over age 75)
- Family history of osteoarthritis
- Occupation that involves daily overworking of joints
- Injury to a joint
- Obesity

- Compression
- Wedging
- Biconcave
- Normal

Osteoporotic vertebral fractures

What is osteoporosis?

Osteoporosis is loss of bone mass due to an imbalance in the bone remodeling cycle. The lack of bone density causes instability and a greater likelihood of broken bones. Bones undergo change on a daily basis. Existing bone tissue is broken down and replaced by new tissue to provide bone mass. Under normal circumstances, the two distinct processes work together to provide a consistent bone mass. As people age, bone remodeling slows, but the tearing down of bone tissue continues at the same pace. The imbalance eventually creates a net loss in bone mass. This natural occurrence is worsened by not providing basic bone-building nutrients to the body such as calcium, proteins and vitamin D.

Risk factors for OP

- Post-menopausal women
- Ethnicity (Caucasian or Asian are at the highest risk)
- Family history of osteoporosis
- Eating disorders such as anorexia or bulimia
- Vigorous exercise program
- Overweight
- Alcoholism
- Thyroid disease
- Prolonged use of the anticoagulant heparin
- Males with reduced testosterone

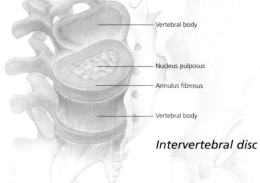

- Vertebral body
- Nucleus pulposus
- Annulus fibrosus
- Vertebral body

Intervertebral disc

Spinal deformities

In the normal spine, the inward curve of the lumbar spine is called **lordotic** and the outward curve of the thoracic spine is called **kyphotic**. Abnormal exaggeration of these curves or sideways curvature of the spine (**scoliosis**) may develop in children or adults with varying degrees of severity.

Lordosis or swayback is an excessive curvature of the lumbar spine. It may be congenital (present at birth) or caused by poor posture, neuromuscular disorders, hip problems, injury or infection. Exercise may stop the progression of the deformity.

Kyphosis is a thoracic spinal deformity that may be congenital or acquired as a result of other conditions, including Scheuermann's disease, which causes excessive curvature of the thoracic vertebrae. Symptoms may include unequal shoulder height, back pain, forward bending of the head and other complications. It may be treated with bracing or surgery.

Scoliosis commonly affects adolescents, particularly girls. Symptoms include side-to-side curvature of the upper or lower spine, unequal shoulder and hip height, sore or stiff back and other complications. Many cases remain mild. More severe cases may be treated by a brace or surgical bonding of the vertebrae.

©Scientific Publishing Ltd., Elk Grove Village, IL, USA
#1051

PLATE 8

Osteoarthritis

Axial skeleton
Appendicular skeleton

What is arthritis?

Arthritis is a general term used to describe any process that causes joint damage. There are more than 100 different types of arthritis. Most involve swelling, tenderness and pain in various joints of the body. **Osteoarthritis** is the most common type of arthritis. Other types include **Rheumatoid Arthritis** and **gout**.

What is osteoarthritis?

Osteoarthritis (OA), also called **degenerative joint disease** or **osteoarthrosis**, has existed for centuries, occurring in many animals as well as humans. Currently, over 20 million people in the U.S. have OA. It is characterized by a gradual loss of cartilage and overgrowth of bone, often within only one or a few joints. Unlike other forms of arthritis, OA does not spread to other parts of the body. OA can occur in almost any movable joint, but most commonly affects the spine, hips, knees, hands or feet.

A healthy joint

A **joint** is any location in the body where two bones come together. Ligaments and other flexible structures surrounding a joint hold the bones together and allow for movement. There are several types of joints that permit different degrees of movement. OA usually occurs in **synovial joints**, which are the most freely movable. Specialized structures inside each of these joints provide a protected, shock-absorbing, self-lubricated environment capable of delivering a wide range of precise movements with minimal friction.

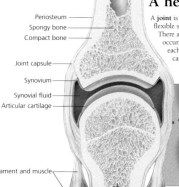

Periosteum
Spongy bone
Compact bone
Joint capsule
Synovium
Synovial fluid
Articular cartilage
Ligament and muscle

Compression   Expansion

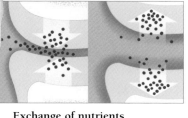

Exchange of nutrients

Unlike cartilage, the synovium is loaded with blood vessels. Synovial fluid secreted by the synovium is rich in nutrients from the blood. Since cartilage is like a dense sponge, repeated compression and expansion during and after joint movement circulates synovial fluid throughout the cartilage, removing waste and delivering necessary nutrients.

Articular cartilage

Inside a joint, articular cartilage covers and protects the surface of each bone. The unique structure of cartilage allows it to bear the brunt of the stress placed on the joint during movement. Different from most tissues, cartilage has no blood vessels or nerves, so inflammation and pain cannot occur. Each surface is smooth and precisely shaped to minimize friction. Large amounts of flexible matrix allow surfaces to constantly adapt as they glide over each other. Collagen fibers more densely packed at the surface distribute stress and prevent harmful substances from entering the cartilage.

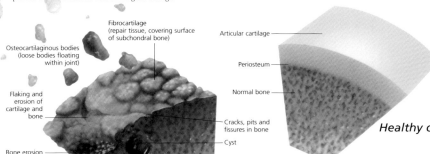

Osteocartilaginous bodies
(loose bodies floating
within joint)

Fibrocartilage
(repair tissue, covering surface
of subchondral bone)

Articular cartilage

Periosteum

Normal bone

Flaking and
erosion of
cartilage and
bone

Cracks, pits and
fissures in bone

Cyst

Bone erosion

Cartilage erosion

Healthy cartilage

Effects of OA

OA is a slow, progressive process that begins at the cellular level within the articular cartilage, probably as early as in the third decade of life. For reasons not currently understood, the collagen meshwork begins to break up, altering the joints' resilience to stress and causing microfractures. Any unsuccessful repair attempts leave the cartilage stiffer, leading to more microfractures. The cartilage surface, once firm and smooth, becomes soft, rough and irregular. As the protective surface wears away, harmful enzymes enter the cartilage and further damage the matrix. Erosion of cartilage continues until bone is exposed.

Once cartilage is damaged, stress from movement is transferred to other structures within the joint. Unlike healthy cartilage, many of these structures hold up very poorly under stress, becoming injured, inflamed and increasingly painful.

Synovial joint with osteoarthritis

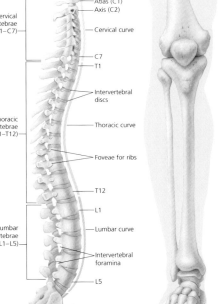

Joint capsule and other
surrounding structures stretch
and weaken as moving the joint
becomes more difficult

Synovium absorbs fragments of
deteriorating cartilage floating in
synovial fluid and becomes thick
and irritated

Synovium attempts to repair
surface of cartilage

Bone changes shape in response
to friction, forming outgrowths
of bone called osteophytes

Instability leads to subluxation,
partial dislocation of the joint

Joint fluid seeps through cracks in
cartilage into the bone marrow, causing
cysts to form

Damaged synovium and decreased
use of joint due to pain can inhibit
nutrient delivery to remaining cartilage

Superficial layer of exposed bone dies

Friction increases as a continuous
film of lubricant fails to form during
movement

Treatment of osteoarthritis

Currently there is no cure for OA, but a great deal can be done to manage and even postpone its effects. Treatment focuses on performing daily activities, managing pain and preventing disability.

Exercise

Daily exercise, the most important element of OA treatment, is essential to maintaining range of motion in a joint, avoiding stiffness, strengthening the structures that surround and protect the joint and delivering nutrients to the cartilage. It is very important to balance exercise with adequate periods of rest.

Medication

Nonmedicinal types of pain relief, such as heat and massage, can be very effective. If medication is needed for pain relief, however, it is important to take the smallest effective dose of the safest effective medication. Acetaminophen is usually tried first because it has very few side effects. NSAIDs (nonsteroidal anti-inflammatory drugs), such as aspirin and ibuprofen, can be tried if acetaminophen is ineffective. In extreme cases, cortisone can be injected directly into a joint to relieve pain. However, cortisone and most NSAIDs can cause more serious side effects.

Surgery

When all other methods of pain relief have failed, surgery might become an option. Surgery can range from modifying a joint to completely replacing a joint with a prosthetic one.

Joints affected by osteoarthritis

■ Most common
■ Less common

Temporomandibular joint (TMJ)

What causes osteoarthritis?

OA occurs whenever articular cartilage stops functioning properly and bones come in contact with each other. Most often, the exact cause of cartilage deterioration is unknown. In primary osteoarthritis, cartilage responds abnormally to many years of normal wear and tear. In secondary osteoarthritis, cartilage is damaged by trauma or disease. Causes of secondary OA include sports injuries, repetitive use of a joint, developmental joint abnormalities, infections, metabolic disorders, endocrine disorders and other diseases.

Shoulder
Elbow
Hip
Wrist
Knee
Ankle

First carpometacarpal joints
Proximal interphalangeal joints (Bouchard's nodes)
Distal interphalangeal joints (Heberden's nodes)
First metatarsophalangeal joint

Vertebrae (T10-12, L1) with osteoarthritis
(Lateral view)

Changes in joint shape cause painful compression of surrounding nerves

Vertebral column
(Lateral view)

Atlas (C1)
Axis (C2)
Cervical vertebrae (C1–C7)
Cervical curve
C7
T1
Intervertebral discs
Thoracic vertebrae (T1–T12)
Thoracic curve
Foveae for ribs
T12
L1
Lumbar vertebrae (L1–L5)
Lumbar curve
Intervertebral foramina
L5
Sacrum
Pelvic curve
Coccyx

Symptoms of OA

- Pain in a joint during or after use, relieved by rest
- Stiffness in a joint following inactivity, relieved by movement
- Discomfort in a joint before or after a change in the weather
- Warmth and tenderness (painful to the touch) in a joint
- Decreased flexibility in a joint
- Enlarged joint
- Crepitus (grating or popping sounds) during joint movement
- Slow progression to constant, more severe pain

Risk factors for OA

- Age (risk increases with age, affecting almost everyone over age 75)
- Family history of osteoarthritis
- Occupation that involves daily overworking of joints
- Injury to a joint
- Obesity

Take control of your arthritis

- Educate yourself about osteoarthritis
- Modify your living space to decrease strain on affected joints; for example, move frequently used items off high shelves
- Keep a daily record of activities, rest, medication and any pain experienced to help determine what works best for you
- Work with your healthcare provider to determine an appropriate balance of exercise and rest
- Exercise every day according to your healthcare provider's instructions
- Discuss potential side effects of medication with your healthcare provider
- If necessary, use splints or braces to support affected joints
- Shed extra weight to decrease strain on your joints

Remember, inactivity can make osteoarthritis worse.

©Scientific Publishing Ltd., Elk Grove Village, IL USA
#1050

PLATE 9

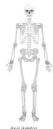

Axial skeleton
Appendicular skeleton

Understanding Rheumatoid Arthritis

What is arthritis?

The term arthritis refers to a number of disease processes that cause damage to the joints and commonly involve swelling, tenderness, stiffness and pain. There are more than 100 different types of arthritis, including **osteoarthritis** and **rheumatoid Arthritis**.

What is Rheumatoid Arthritis?

Rheumatoid Arthritis (RA) is a chronic, systemic inflammatory disease primarily affecting the synovial joints. For unknown reasons, the immune system attacks the linings (**synovial membranes**) of joints such as the hands, wrists, feet and knees.

The joints often become chronically inflamed, leading to swelling, pain, stiffness and changes in joint mobility and function.

Fluid and inflammatory cells accumulate in the synovium to produce **pannus**, an invasive tissue that covers the surface of the joint's articular cartilage and erodes the cartilage, bone, ligaments and tendons.

Continuing inflammation of the synovium can eventually cause irreversible damage to the bones of the joint.

What causes Rheumatoid Arthritis?

The cause of rheumatoid arthritis remains unknown, but is thought to be linked to a combination of genetic and environmental factors triggering an abnormal or defective immune response.

Factors under investigation for their potential role in the development of RA include autoantibodies such as rheumatoid factor (RF), proinflammatory substances known as cytokines (including tumor necrosis factor, or TNF) and possible infectious microorganisms.

Research into the causes and development of RA continues to evolve and provides new direction for treatment strategies.

Hand with early Rheumatoid Arthritis

Joint pathology in Rheumatoid Arthritis

The inflammatory process is a function of the immune system, which defends the body against trauma or infection or invasion by foreign substances.

In RA, the immune system attacks the tissues that cushion and surround the joints. An influx of inflammatory cells into the synovial membrane in early RA causes thickening of the joint lining and subsequent soft tissue swelling.

Over time, this destructive process leads to cartilage and soft tissue damage, pannus formation, fibrosis, bone erosion and joint deformities.

This figure shows swelling and acute inflammation of the synovial membrane. Pannus formation caused by synovitis (chronic inflammation) leads to bone erosions, which may occur rapidly during the first two years of the disease.

Bone
Cartilage
Pannus
Inflamed synovial membrane
'Bare' bone erosion
Joint capsule

Joint with early RA

Neck
Shoulder
Elbow
Hip
Wrist
Fingers
Knee
Ankle
Foot

Joints most often affected by Rheumatoid Arthritis

Signs and symptoms of RA

Symptoms listed below may appear suddenly or develop over time.

- Joint pain, warmth, swelling or tenderness
- Joints affected in symmetrical pattern; e.g., both wrists, both shoulders
- Joint stiffness, usually lasting longer than an hour after awakening or inactivity
- A feeling of generalized stiffness
- Fatigue and weakness
- Low-grade fever
- Anorexia and weight loss
- Limited joint mobility
- Subcutaneous nodules
- Evidence of joint or bone erosion

Some risk factors for RA

- Women are two to three times more likely to develop RA than men
- Family history of rheumatoid arthritis

Early RA

Early-stage symptoms can include pain, weakness, weight loss, low-grade fever and generalized fatigue.

Soft tissue stiffness, swelling and tenderness may be present in the joints of the fingers, wrists, knees and feet, caused by inflammation of the synovial membranes (synovitis).

Subcutaneous nodules may also be present.

Current diagnostic techniques for identifying and monitoring early RA include physical exams, blood tests and x-rays as well as MRI (magnetic resonance imaging), which has been shown to be effective for detecting early erosion of bare bone (areas lacking protective cartilage, such as joint margins).

Hand with moderate Rheumatoid Arthritis

Joint erosion

This figure shows continuing bone erosion, which may lead to increased loss of joint space.

Loss of joint space
Bone erosion
Mild osteoporosis

Joint with moderate RA

Moderate RA

Chronic synovitis and pannus formation cause continuing bone and soft tissue erosion, leading to joint instability and deformity. Narrowing of the spaces between the bones can be seen in the hand and other joints. The inflammatory process causes damage to the joint's supporting structures, stretching tendons and ligaments and creating progressive malalignment or partial joint dislocation (**subluxation**).

TNF and TNF receptors

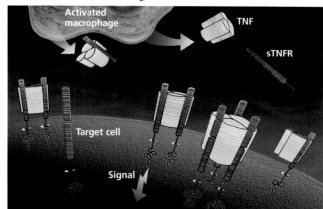

Activated macrophage
TNF
sTNFR
Target cell
Signal

Understanding inflammation

As part of the chronic inflammatory response seen in RA, a destructive layer of cells called pannus forms and covers the articular surfaces of the joints, gradually eroding cartilage and bone.

The inflammatory properties of the rheumatoid synovial fluid in joints with RA also contribute to overall inflammation and damage to the surrounding tissues.

One important inflammatory cytokine found in synovial fluid is called tumor necrosis factor (TNF), which has been associated with the production of enzymes that attack the cartilage. TNF also stimulates pannus cells to produce other destructive substances.

Hand with severe Rheumatoid Arthritis

Ulnar deviation
Boutonnière deformity
Swan-neck deformity

Severe RA

Untreated, persistent synovitis with erosion of bone, cartilage and soft tissues eventually causes significant joint damage and deformity. Among the joints severely affected are the **knuckles** (metacarpophalangeal or **MCP** joints) and **finger joints** (proximal and distal interphalangeal or **PIP** and **DIP** joints). Disabling hand conditions include **Swan-neck** deformity (hyperextension of the PIP joint while the DIP joint is flexed) and **Boutonnière deformity** (hyperextension of the DIP joint with flexion of the PIP joint). In some patients, small **subcutaneous nodules** may be present under the skin of the forearms and elbow.

This figure shows ulnar deviation in severe RA. The fingers have been forced away from the thumb by uneven bone erosion and tendon damage.

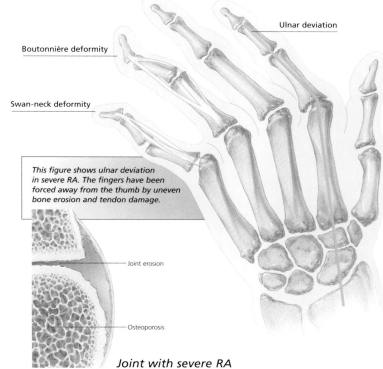

Joint erosion
Osteoporosis

Joint with severe RA

Treatment of Rheumatoid Arthritis

While there is presently no cure for RA, treatments are available to help manage and even inhibit its effects. DMARD (disease-modifying antirheumatic drug) therapy is prescribed to relieve pain and inflammation. Early treatment with DMARDs may also increase the window of opportunity to help prevent structural joint damage.

Exercise

Daily exercise often helps to maintain range of motion in joints, avoid stiffness and strengthen the structures that surround and protect joints. It is important to balance exercise with adequate periods of rest.

Medication

The object of most drug therapies is to relieve pain, reduce inflammation, help inhibit the progression of joint damage and deformity, preserve movement and control systemic involvement of the disease. Nonmedical types of pain relief, such as heat and massage, can be helpful.

Surgery

When all other methods of pain relief have failed, surgery may become an option. Surgery can range from tendon repair to complete joint replacement.

Taking Control of Your Rheumatoid Arthritis

- Educate yourself about rheumatoid arthritis
- Work with your healthcare provider to determine an appropriate balance of exercise and rest
- Discuss appropriate medical treatments with your health-care provider
- Follow a nutritious and healthy diet, including plant and fish oils to help reduce inflammation
- Shed extra weight to decrease strain on your joints

Remember, early detection and effective therapy are important in successful management of your rheumatoid arthritis.

©Scientific Publishing Ltd., Elk Grove Village, IL USA
#1152

PLATE 10

Understanding Osteoporosis

Axial skeleton
Appendicular skeleton.

What is osteoporosis?

Osteoporosis is loss of bone mass due to an imbalance in the bone remodeling cycle. The lack of bone density causes instability and a greater likelihood of broken bones. Bones undergo a change on a daily basis. Existing bone tissue is broken down and replaced by new tissue to provide bone mass. Under normal circumstances, the two distinct processes work together to provide a consistent bone mass. As people age, bone remodeling slows, but the tearing down of bone tissue continues at the same pace. The imbalance eventually creates a net loss in bone mass. This natural occurrence is worsened by not providing basic bone-building nutrients to the body such as calcium, proteins and vitamin D.

What causes osteoporosis?

Osteoporosis is a common side effect of aging. Its severity is a function of risk factors, such as gender, genetics, diet and lifestyle. The highest risk group is post-menopausal women, who have lower levels of estrogen. (Estrogen carries calcium to bone tissue.) Younger women who experience amenorrhea (lack of menstrual periods) are at greater risk as well. Women who exercise excessively or have an eating disorder such as anorexia nervosa can develop osteoporosis earlier in life. At any age, whether male or female, a calcium-deficient diet would tend to increase the risk of osteoporosis.

Areas most affected by osteoporosis

Bone density changes

Bone tissue density changes continually, because of the cycle of replenishing old with new tissue. From birth through adolescence, the cycle places heavier emphasis on building new tissue, resulting in a net increase in bone mass. As people reach their 20's and 30's, bone tissue is maintained at a healthy level because of the normal tissue formation cycle and a healthy diet and lifestyle. As women reach menopause, lack of estrogen production hinders the building of bone tissue. Men are affected to a lesser extent, but do experience net bone tissue loss.

Osteoporotic vertebral fractures

Compression
Wedging
Biconcave
Normal

The bone growth cycle
Formation and restoration

The process of bone remodeling continuously occurs throughout a person's life. Old bone is broken down and replaced by new bone. In osteoporosis, more bone is destroyed than is created, which results in a loss of bone mass.

1. The cycle begins on the trabecular plates (mature bone)
2. Activation of dormant cells, called osteoclast precursors, into bone breakdown cells called osteoclasts
3. Osteoclasts dissolve old bone and dig microscopic cavities
4. Bone-forming osteoblasts are attracted to the cavities, and begin filling them with a collagen matrix
5. Calcium & phosphorous crystals are added to the collagen matrix to strengthen & harden the bone

Colle's fracture

Radius

Sites of hip fractures

Femoral neck
Trochanteric

Osteoporotic bone

Cortical (compact) bone
Osteon (Haversian system)
Trabecular (spongy) bone
Medullary cavity

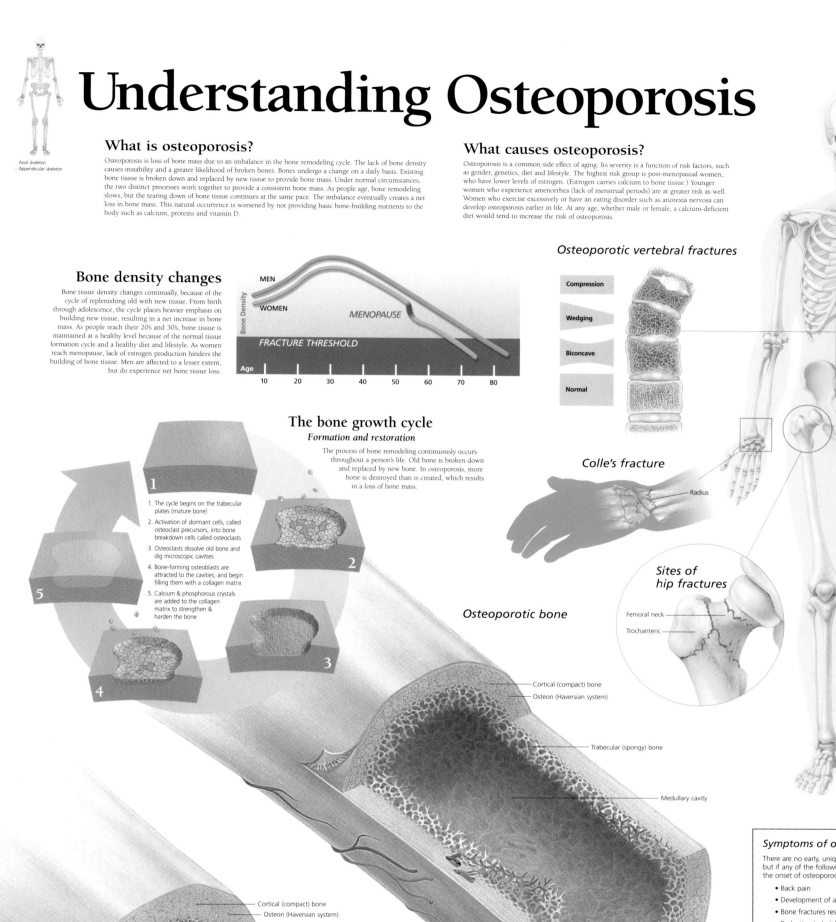

Cortical (compact) bone
Osteon (Haversian system)
Periosteum
Trabecular (spongy) bone
Medullary cavity

Normal bone

Symptoms of osteoporosis

There are no early, unique or distinctive signs of osteoporosis, but if any of the following occur, a physician might diagnose the onset of osteoporosis.

- Back pain
- Development of a hunched back and abdominal protrusion
- Bone fractures resulting from an apparent minor trauma
- Reduction in height

Risk factors

- Post-menopause estrogen deficiency
- Ethnicity (Caucasian or Asian groups are at the highest risk)
- Family history of osteoporosis
- Eating disorders such as anorexia or bulimia
- Vigorous exercise program
- Overweight
- Alcoholism
- Thyroid disease
- Prolonged use of the anticoagulant heparin
- Males with reduced testosterone

Taking control of your osteoporosis

- Educate yourself about osteoporosis — it's easier to prevent than treat
- Maintain a calcium-rich diet — 1,000 mg to 1,200 mg per day
- Include vitamin D supplements in your diet
- Follow a reasonable exercise program
- Maintain a balanced posture to limit stress on the spine
- Take hormone replacement therapy, if directed by a physician

Effects of osteoporosis (progressive spinal deformity)

©Scientific Publishing Ltd., Elk Grove Village, IL. USA
#1153

PLATE 11

The Respiratory System

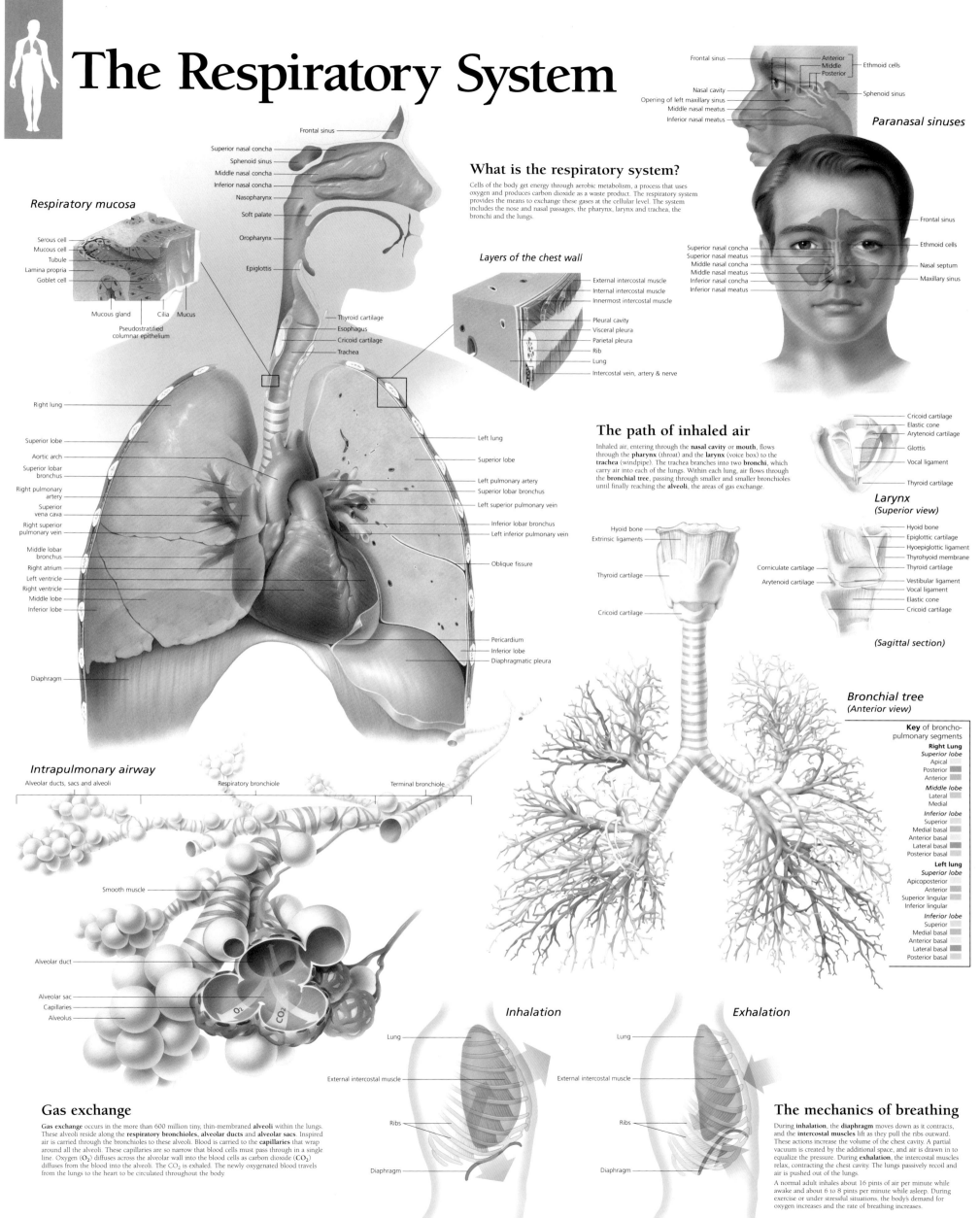

Respiratory mucosa

Serous cell
Mucous cell
Tubule
Lamina propria
Goblet cell

Mucous gland
Cilia
Mucus
Pseudostratified columnar epithelium

Frontal sinus
Superior nasal concha
Sphenoid sinus
Middle nasal concha
Inferior nasal concha
Nasopharynx
Soft palate
Oropharynx
Epiglottis

Right lung

Superior lobe
Aortic arch
Superior lobar bronchus
Right pulmonary artery
Superior vena cava
Right superior pulmonary vein
Middle lobar bronchus
Right atrium
Left ventricle
Right ventricle
Middle lobe
Inferior lobe

Diaphragm

Thyroid cartilage
Esophagus
Cricoid cartilage
Trachea

Left lung
Superior lobe
Left pulmonary artery
Superior lobar bronchus
Left superior pulmonary vein
Inferior lobar bronchus
Left inferior pulmonary vein
Oblique fissure
Pericardium
Inferior lobe
Diaphragmatic pleura

What is the respiratory system?

Cells of the body get energy through aerobic metabolism, a process that uses oxygen and produces carbon dioxide as a waste product. The respiratory system provides the means to exchange these gases at the cellular level. The system includes the nose and nasal passages, the pharynx, larynx and trachea, the bronchi and the lungs.

Layers of the chest wall

External intercostal muscle
Internal intercostal muscle
Innermost intercostal muscle
Pleural cavity
Visceral pleura
Parietal pleura
Rib
Lung
Intercostal vein, artery & nerve

The path of inhaled air

Inhaled air, entering through the **nasal cavity** or **mouth**, flows through the **pharynx** (throat) and the **larynx** (voice box) to the **trachea** (windpipe). The trachea branches into two **bronchi**, which carry air into each of the lungs. Within each lung, air flows through the **bronchial tree**, passing through smaller and smaller bronchioles until finally reaching the **alveoli**, the areas of gas exchange.

Hyoid bone
Extrinsic ligaments
Thyroid cartilage
Cricoid cartilage

Paranasal sinuses

Frontal sinus
Anterior
Middle
Posterior
Ethmoid cells
Nasal cavity
Opening of left maxillary sinus
Middle nasal meatus
Inferior nasal meatus
Sphenoid sinus

Frontal sinus
Superior nasal concha
Superior nasal meatus
Middle nasal concha
Middle nasal meatus
Inferior nasal concha
Inferior nasal meatus
Ethmoid cells
Nasal septum
Maxillary sinus

Cricoid cartilage
Elastic cone
Arytenoid cartilage
Glottis
Vocal ligament
Thyroid cartilage

Larynx
(Superior view)

Hyoid bone
Epiglottic cartilage
Hyoepiglottic ligament
Thyrohyoid membrane
Thyroid cartilage
Vestibular ligament
Vocal ligament
Elastic cone
Cricoid cartilage

Corniculate cartilage
Arytenoid cartilage

(Sagittal section)

Bronchial tree
(Anterior view)

Key of broncho-pulmonary segments

Right Lung
Superior lobe
Apical
Posterior
Anterior
Middle lobe
Lateral
Medial
Inferior lobe
Superior
Medial basal
Anterior basal
Lateral basal
Posterior basal

Left lung
Superior lobe
Apicoposterior
Anterior
Superior lingular
Inferior lingular
Inferior lobe
Superior
Medial basal
Anterior basal
Lateral basal
Posterior basal

Intrapulmonary airway

Alveolar ducts, sacs and alveoli
Respiratory bronchiole
Terminal bronchiole

Smooth muscle
Alveolar duct
Alveolar sac
Capillaries
Alveolus

O_2
CO_2

Gas exchange

Gas exchange occurs in the more than 600 million tiny, thin-membraned **alveoli** within the lungs. These alveoli reside along the **respiratory bronchioles, alveolar ducts** and **alveolar sacs**. Inspired air is carried through the bronchioles to these alveoli. Blood is carried to the **capillaries** that wrap around all the alveoli. These capillaries are so narrow that blood cells must pass through in a single line. Oxygen (O_2) diffuses across the alveolar wall into the blood cells as carbon dioxide (CO_2) diffuses from the blood into the alveoli. The CO_2 is exhaled. The newly oxygenated blood travels from the lungs to the heart to be circulated throughout the body.

Inhalation

Lung
External intercostal muscle
Ribs
Diaphragm

Exhalation

Lung
External intercostal muscle
Ribs
Diaphragm

The mechanics of breathing

During **inhalation**, the **diaphragm** moves down as it contracts, and the **intercostal muscles** lift as they pull the ribs outward. These actions increase the volume of the chest cavity. A partial vacuum is created by the additional space, and air is drawn in to equalize the pressure. During **exhalation**, the intercostal muscles relax, contracting the chest cavity. The lungs passively recoil and air is pushed out of the lungs.

A normal adult inhales about 16 pints of air per minute while awake and about 6 to 8 pints per minute while asleep. During exercise or under stressful situations, the body's demand for oxygen increases and the rate of breathing increases.

©Scientific Publishing Ltd., Elk Grove Village, IL, USA
#1300

PLATE 12

Understanding Asthma

What is asthma?

Asthma is a chronic inflammatory disease of the lungs in which the **bronchial airways** periodically and temporarily narrow in response to stimuli. Normally, these airways narrow to prevent harmful substances from entering the lungs, but in asthmatics the airways narrow too easily, too much, and in response to things that wouldn't ordinarily cause a reaction. This narrowing makes breathing difficult. Asthma is a common disease affecting approximately 10% of children and 5% of adults.

Breathing & gas exchange

Breathing provides the **oxygen** (O_2) we need to live and expels the **carbon dioxide** (CO_2) we need to eliminate. A complete breath consists of taking air into the **lungs** (**inspiration**) and then expelling it (**expiration**). Inspired air travels through the lungs' branching network of progressively smaller airways (**bronchi and bronchioles**) until reaching the more than 300 million tiny, thin-membraned **alveoli** where **gas exchange** occurs. Gas exchange of O_2 and CO_2 occurs through the walls of the alveoli and the walls of the blood-carrying capillaries that envelop the alveoli.

Measuring air flow

Measuring air movement out of your lungs can be helpful in understanding and managing asthma. The **peak expiratory flow rate (PEFR)** is a measurement of the force and speed of air blown out of the lungs and can be measured at home with an easy-to-use **peak flow meter**. By recording your results on a **peak flow chart** and comparing them to your personal best score (the highest score over a 2-3 week period while your asthma is under good control), you can gauge how your asthma is doing, what makes it worse, how effective your medications are, and when to seek emergency help.

Peak flow chart

Dips below safe range mean asthma is worsening and needs special attention

Safe range within 80% of personal best

Allergies & asthma

Allergies play a significant role for many asthmatics: about 80% of children and 50% of adults with asthma have allergies. Allergies are the result of the body's **immune system** perceiving a harmless foreign substance (**allergen**) as dangerous. In asthmatics, allergic reactions can trigger an asthma attack when common allergens, such as pollens, mold spores, dust mites, cockroach particles, and animal danders are inhaled into the lungs.

People with allergies make large quantities of certain **antibodies** (IgE) against specific allergens. These antibodies are located on the surface of **mast cells**—immune system cells abundant in the lining of the nose and lungs. When these antibodies are exposed to the allergen, mast cells are triggered to release chemicals called **mediators** (histamine) into the surrounding area. These chemicals irritate the tissues and cause allergic symptoms, such as runny eyes and nose, but also the immediate response of an asthma attack—bronchospasm and mucus production.

Close-up of airway wall

Antigen
Mucus
Cilia
Antibodies attached to mast cell
Epithelial cells
Goblet cells

Airway with allergies

Immediate reaction (in minutes)

Antigen attaches to antibodies
Mast cell releases mediators
Bronchospasm
Mucus
Open airway
Inflammatory cells thicken bronchial wall

Late reaction (in hours)

After the immediate allergic reaction, the chemicals released by mast cells also may produce a late (delayed) reaction in the airways. Inflammatory cells are attracted to the area and cause the airways to again narrow and swell. This late reaction can last much longer than the initial reaction and can be more dangerous.

Inflammation

It was once believed that asthma was simply episodes of bronchospasm by the smooth muscle. Today, the focus is **inflammation**. Inflammation is thought to play a critical role in airway obstruction and trigger sensitivity. Controlling inflammation has become a central objective of asthma therapy. Reducing the underlying inflammation is believed to reduce spontaneous bronchospasm and possibly prevent long-term damage to the airways. Better control of inflammation is essential to better long-term and short-term control of asthma.

Children & asthma

Asthma is the most common chronic illness in children. The number of cases is increasing and many are believed yet undiagnosed. Childhood respiratory asthma is thought to be genetically predisposed and is closely associated with allergies. Frequent respiratory infections and coughing, including coughing after exercise, recurrent nighttime coughing and rattly coughs of infants, should be evaluated for asthma. Children with asthma should be tested for allergies and once identified, avoid triggers, such as second-hand smoke. The ultimate goals of treating children are to rid them of symptoms, restore normal lung function and allow them to lead a *full and active* life.

What causes asthma?

The exact cause of asthma is not known, but it is believed that there are hereditary and environmental factors involved. Once someone has asthma, there are factors that can cause or "**trigger**" an asthma attack. Identifying and avoiding triggers can help control asthma and play an important role in its treatment.

Common triggers

- Allergens - *dust, molds, pollens, animal danders, cockroach particles*
- Exercise - *vigorous activities like running*
- Cold damp air
- Time of day - *early morning hours*
- Respiratory tract infections - *viruses*
- Emotions - *laughing, crying, stress*
- Drugs - *especially aspirin and other NSAIDs*
- Occupational factors - *triggers at work*
- Irritants - *tobacco smoke, household cleaners, etc.*
- Ozone

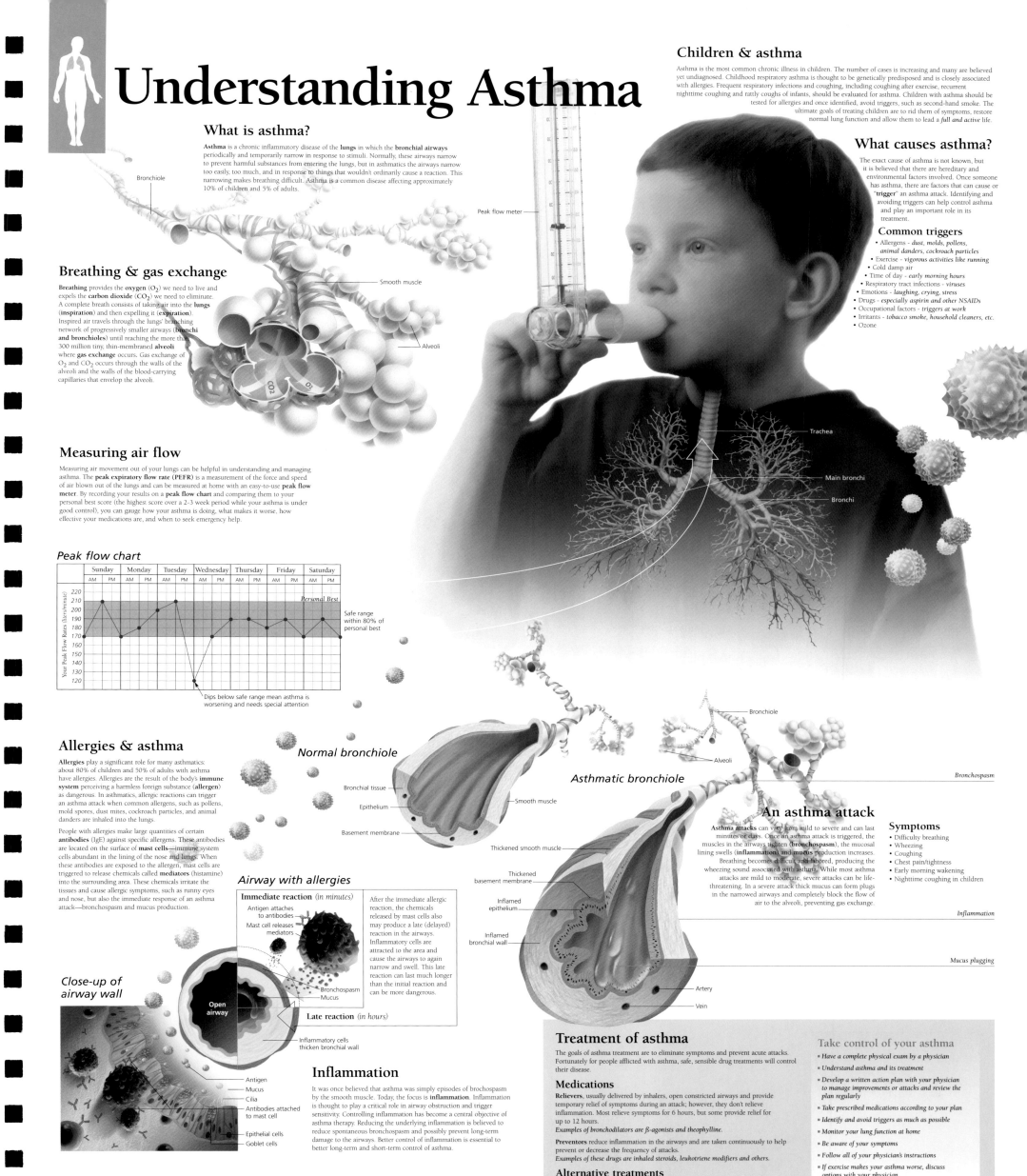

Peak flow meter
Bronchiole
Smooth muscle
Alveoli
Trachea
Main bronchi
Bronchi

Normal bronchiole
Bronchial tissue
Epithelium
Basement membrane
Smooth muscle

Asthmatic bronchiole
Thickened smooth muscle
Thickened basement membrane
Inflamed epithelium
Inflamed bronchial wall
Artery
Vein
Bronchiole
Alveoli
Bronchospasm
Inflammation
Mucus plugging

An asthma attack

Asthma attacks can vary from mild to severe and can last minutes or days. Once an asthma attack is triggered, the muscles in the airways tighten (**bronchospasm**), the mucosal lining swells (**inflammation**) and mucus production increases. Breathing becomes difficult and labored, producing the wheezing sound associated with asthma. While most asthma attacks are mild to moderate, severe attacks can be life-threatening. In a severe attack thick mucus can form plugs in the narrowed airways and completely block the flow of air to the alveoli, preventing gas exchange.

Symptoms

- Difficulty breathing
- Wheezing
- Coughing
- Chest pain/tightness
- Early morning wakening
- Nighttime coughing in children

Treatment of asthma

The goals of asthma treatment are to eliminate symptoms and prevent acute attacks. Fortunately for people afflicted with asthma, safe, sensible drug treatments will control their disease.

Medications

Relievers, usually delivered by inhalers, open constricted airways and provide temporary relief of symptoms during an attack; however, they don't relieve inflammation. Most relieve symptoms for 6 hours, but some provide relief for up to 12 hours.
Examples of bronchodilators are ß-agonists and theophylline.

Preventors reduce inflammation in the airways and are taken continuously to help prevent or decrease the frequency of attacks.
Examples of these drugs are inhaled steroids, leukotriene modifiers and others.

Alternative treatments

Yoga, hypnosis and acupuncture may help relieve anxiety and hyperventilation that can make asthma attacks worse.

Take control of your asthma

- *Have a complete physical exam by a physician*
- *Understand asthma and its treatment*
- *Develop a written action plan with your physician to manage improvements or attacks and review the plan regularly*
- *Take prescribed medications according to your plan*
- *Identify and avoid triggers as much as possible*
- *Monitor your lung function at home*
- *Be aware of your symptoms*
- *Follow all of your physician's instructions*
- *If exercise makes your asthma worse, discuss options with your physician*

©Scientific Publishing Ltd., Elk Grove Village, IL. USA
#1350

PLATE 13

COPD
Chronic Obstructive Pulmonary Disease

Chronic Obstructive Pulmonary Disease (COPD)

Chronic Obstructive Pulmonary Disease (COPD), also called chronic obstructive lung disease, is a name used for two related diseases of the respiratory system: **emphysema** and **chronic bronchitis**. In many individuals these diseases occur together, although there may be more symptoms of one than the other. The majority of individuals with COPD have a long history of cigarette smoking.

COPD gradually worsens with time. Initially there may be only a mild shortness of breath and occasional coughing. A chronic cough then develops with a clear, colorless sputum. As the disease progresses, the cough worsens and more effort is needed to get air into and out of the lungs. In later stages, the heart might become affected. Eventually death occurs when the functioning of the lungs and heart is no longer adequate to supply oxygen to the body's organs and tissues.

Approximately 80%–90% of COPD cases are caused by smoking. Air pollution and occupational exposures play a role, especially when combined with cigarette smoking. By the time symptoms of COPD appear—typically cough, shortness of breath and difficulty tolerating exercise—damage has already occurred to your lungs.

Emphysema

Signs & symptoms:
• Shortness of breath
• Chronic, mild cough that may produce sputum
• Weight loss

On average, the lungs contain 300 million elastic air sacs, called **alveoli**, in which oxygen is added to the blood and carbon dioxide is removed from it. Emphysema occurs when there is permanent damage to the alveoli and they lose their natural elasticity, become over stretched and rupture, preventing the lungs from properly functioning. This results in the bloodstream not receiving the required amounts of oxygen.

What causes emphysema?
The normal lung has a unique balance between two chemicals with opposing actions. The elastic fibers allow the lung to expand and contract. When the chemical balance is altered the lungs lose the ability to protect themselves against the destruction of these elastic fibers. Smoking is responsible for the majority (80%-90%) of emphysema cases. Individuals born with a protein deficiency known as alpha-1 antitrypsin (AAT) may develop to an inherited form of emphysema.

Chronic bronchitis

Signs & symptoms:
• Chronic cough that produces mucus
• Shortness of breath

This disorder consists of chronic inflammation and thickening of the lining of the bronchial tubes. Pushing the air through narrowed airways becomes harder and harder. In addition, the inflammation causes the glands of the bronchial tubes to produce excessive amounts of mucus, increasing congestion in the lungs and further inhibiting the ability to breathe. Air flow is hindered and the lungs are endangered.

The primary symptom of chronic bronchitis (as distinct from emphysema) is a chronic cough that produces a large amount of mucus and has persisted for at least 3 months of the year for more than 2 consecutive years. Once the bronchial tubes have been irritated over a long period of time, excessive mucus is produced constantly.

What causes chronic bronchitis?
In addition to smoking, higher rates of chronic bronchitis are found among coal miners, grain handlers, metal molders and other workers exposed to dust.

Breathing

All humans need **oxygen** to burn nutrients, which release the energy we need to live. Through breathing, our respiratory system provides this needed oxygen and expels the **carbon dioxide** we need to eliminate. A complete breath includes taking air into the lungs (**inspiration**) and then expelling it (**expiration**). A normal adult inhales about 16 pints of air per minute while awake and about 6 to 8 pints per minute while asleep. During exercise or under stressful situations, the body's demand for oxygen increases and the rate of breathing increases.

Enlargement of lung tissue:
(*left*) From a normal lung.
(*right*) From a lung with emphysema.
In emphysema, lung tissue is destroyed, resulting in fewer and larger alveoli.

Labels (respiratory diagram): Frontal sinus, Superior nasal concha, Sphenoid sinus, Middle nasal concha, Inferior nasal concha, Nasopharynx, Soft palate, Oropharynx, Epiglottis, Thyroid cartilage, Esophagus, Cricoid cartilage, Trachea

Labels (alveoli): Normal alveoli, Damaged alveoli, Alveolar sac, Capillaries, Alveolus, O_2, CO_2

Labels (bronchitis diagram): Bronchospasm, Mucus, Open airway, Thickened bronchial wall

Labels (bronchial cross-section): Thickened smooth muscle, Thickened basement membrane, Inflamed epithelium, Inflamed bronchial wall, Mucus, Vein, Artery

Taking control of your COPD

▪ Don't smoke.
▪ Avoid exposures to dusts and fumes.
▪ Avoid air pollution and cigarette smoke.
  ▪ Limit activities during air pollution and ozone alerts.
    ▪ Avoid excessive heat, cold and very high altitudes. (Most COPD patients can travel on commercial airlines with pressurized cabins.)
    ▪ Limit exposure to people with respiratory infections, colds and the flu.
    ▪ Maintain a normal weight. Being over- or underweight can worsen the conditions of COPD.
  ▪ Drink lots of fluids to loosen sputum so it can be easily coughed up.
▪ Follow a nutritious, well-balanced diet.
▪ Follow all of your physician's instructions.
▪ Take prescribed medications as part of your daily routine.
▪ Don't take other people's medications.
▪ Consult your physician about an appropriate exercise plan and follow it.

Effective control of COPD can prevent most of its complications.

How is Chronic Obstructive Pulmonary Disease detected?

There are no accurate methods to predict an individual's chance of developing COPD. None of the current ways to diagnose COPD detects the disease before irreversible lung damage occurs.

Pulmonary function tests
Pulmonary function tests (PFTs) are used to determine lung characteristics and capabilities. These tests include:
• **Total lung capacity:** the amount of air the lungs can hold.
• **Forced expiratory volume:** how quickly air moves in and out of the lungs.
• **Arterial blood gas - pulse oximetry:** how efficiently the lungs transfer oxygen from the air into the blood.
• **Arterial blood gas:** how efficiently the lungs remove carbon dioxide from the blood.
• **X-ray:** in moderate to severe cases, a reasonably accurate diagnosis of COPD can be made with a plain chest x-ray and CAT (computerized axial tomography) scanning.

In most cases, it is necessary to compare the results of several tests in order to make the correct diagnosis, and to repeat some tests at intervals to determine the rate of disease progression or improvement. Test results are compared to values considered healthy for an individual's sex, age, weight, height and race.

How is Chronic Obstructive Pulmonary Disease treated?

Although there is no cure for COPD, the disease can be prevented in many cases. In almost all cases the symptoms can be reduced. Survival of individuals with COPD is closely related to the level of their lung function when they are diagnosed and the rate at which they lose this function. The median survival is about 10 years for those with COPD who have lost approximately two-thirds of their normally expected lung function at diagnosis.

There are a number of treatments which can help individuals with COPD. These treatments can be separated into several categories:
I. **Bronchodilators** help open narrowed bronchus and bronchial tubes.
II. **Anti-inflammatories (steroids)** reduce inflammation of the airway walls.
III. **Continuous oxygen therapy** is recommended for individuals with low blood oxygen levels.
IV. **Lung reduction surgery** removes damaged areas of the lung so it can perform more efficiently.
V. **Transplant surgery** is a highly complex procedure that is considered a viable option only in a select group of individuals.
VI. **Pulmonary rehabilitation programs** can be combined with medical treatment to improve overall physical endurance and sense of well-being.

Medication and exercise

Medications for COPD can be given in several forms. The two most common are inhaled or pill medications. Metered-dose inhalers (MDIs) are a convenient, safe way to deliver medication. Because the medication goes directly to the lungs, smaller doses can be used with minimal side effects. Proper technique in using hand-held inhalers is very important to their effectiveness.

Frequently prescribed medications for COPD patients include:
Bronchodilators to open narrowed airways
Corticosteroids or steroids to reduce inflammation
Antibiotics to fight respiratory infections
Expectorants to loosen and expel mucus secretions from airways
Diuretics to help excrete excess body fluids
Digitalis to strengthen the force of the heartbeat
Other drugs may include tranquilizers, pain killers and cough suppressants.

After smoking cessation, exercise is important to the nonmedical treatment of COPD. Exercise builds and maintains strength, maintains flexibility of the bones and joints and builds stamina to increase the amount of activity possible for a COPD patient. A physician, respiratory therapist or physical therapist should always be consulted before setting up a specific exercise program.

©Scientific Publishing Ltd., Elk Grove Village, IL USA
#1351

PLATE 14

The Effects of Smoking

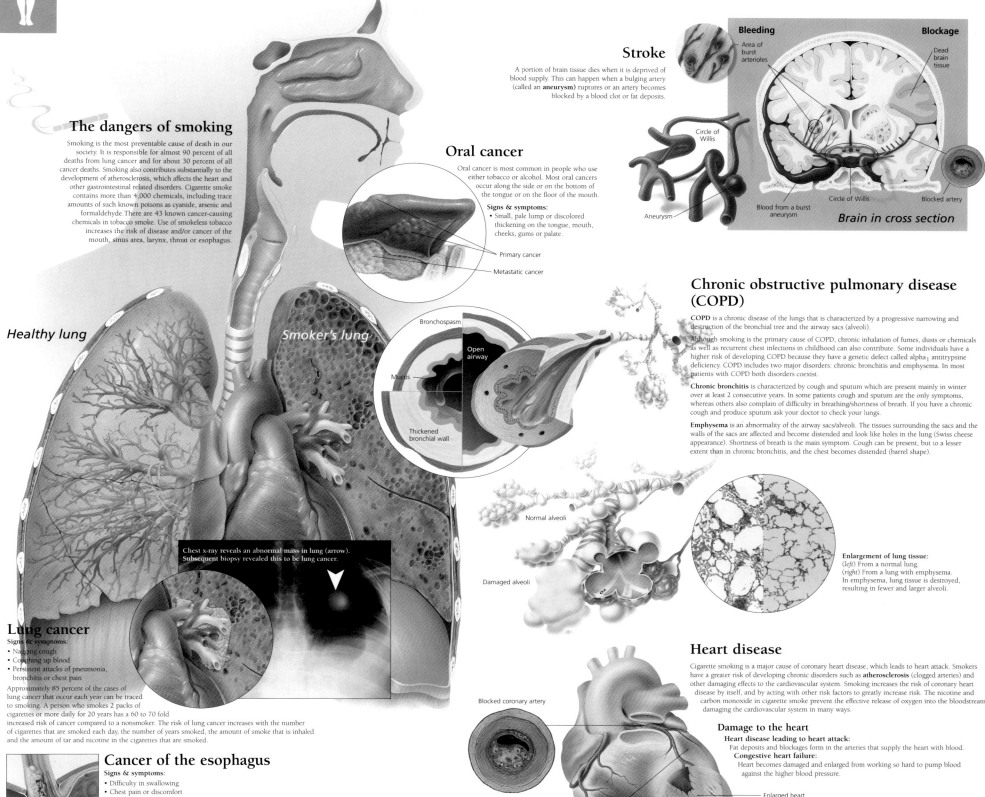

The dangers of smoking

Smoking is the most preventable cause of death in our society. It is responsible for almost 90 percent of all deaths from lung cancer and for about 30 percent of all cancer deaths. Smoking also contributes substantially to the development of atherosclerosis, which affects the heart and other gastrointestinal related disorders. Cigarette smoke contains more than 4,000 chemicals, including trace amounts of such known poisons as cyanide, arsenic and formaldehyde. There are 43 known cancer-causing chemicals in tobacco smoke. Use of smokeless tobacco increases the risk of disease and/or cancer of the mouth, sinus area, larynx, throat or esophagus.

Healthy lung

Smoker's lung

Chest x-ray reveals an abnormal mass in lung (arrow). Subsequent biopsy revealed this to be lung cancer.

Lung cancer
Signs & symptoms:
• Nagging cough
• Coughing up blood
• Persistent attacks of pneumonia, bronchitis or chest pain

Approximately 85 percent of the cases of lung cancer that occur each year can be traced to smoking. A person who smokes 2 packs of cigarettes or more daily for 20 years has a 60 to 70 fold increased risk of cancer compared to a nonsmoker. The risk of lung cancer increases with the number of cigarettes that are smoked each day, the number of years smoked, the amount of smoke that is inhaled and the amount of tar and nicotine in the cigarettes that are smoked.

Cancer of the esophagus
Signs & symptoms:
• Difficulty in swallowing
• Chest pain or discomfort
• Weight loss

Tobacco use can cause esophageal cancer by damaging the structure of cells that line the inside of the esophagus. The longer a person uses tobacco the higher the risk.

Gastric ulcers
Signs & symptoms:
• Gnawing or burning pain in the abdomen
• Nausea, vomiting
• Loss of appetite and weight

The effects of prolonged exposure to smoking contribute to a buildup of stomach acids that erode the protective lining of the stomach. Gnawing or burning pain in the abdomen between the breastbone and the navel is the most common symptom, often occurring between meals and in the early hours of the morning. The pain can last for anything from a few minutes to a few hours and may be relieved by eating or by taking antacids. Smoking slows the healing of existing ulcers and also contributes to ulcer recurrence.

Stomach cancer

Stomach cancer does not usually produce symptoms in the early stages of its development. It is known that stomach cancer can develop from the symptoms of gastric ulcers, and that excessive smoking increases a person's risk of developing the disease.

Other GI diseases

Smoking has also been indicated as a risk factor for Crohn's disease and possibly gallstones.

Bladder cancer
Signs & symptoms:
• Blood in the urine
• Pelvic pain
• Difficulty in voiding urine

Bladder cancer is prevalent among smokers over 40. Men are 4 times more likely than women to get bladder cancer. The presence of blood in the urine without pain or discomfort is the most common early symptom.

Oral cancer

Oral cancer is most common in people who use either tobacco or alcohol. Most oral cancers occur along the side or on the bottom of the tongue or on the floor of the mouth.

Signs & symptoms:
• Small, pale lump or discolored thickening on the tongue, mouth, cheeks, gums or palate.

Primary cancer

Metastatic cancer

Stroke

A portion of brain tissue dies when it is deprived of blood supply. This can happen when a bulging artery (called an **aneurysm**) ruptures or an artery becomes blocked by a blood clot or fat deposits.

Bleeding

Blockage

Area of burst arterioles

Dead brain tissue

Circle of Willis

Aneurysm

Blood from a burst aneurysm

Circle of Willis

Blocked artery

Brain in cross section

Chronic obstructive pulmonary disease (COPD)

COPD is a chronic disease of the lungs that is characterized by a progressive narrowing and destruction of the bronchial tree and the airway sacs (alveoli).

Although smoking is the primary cause of COPD, chronic inhalation of fumes, dusts or chemicals as well as recurrent chest infections in childhood can also contribute. Some individuals have a higher risk of developing COPD because they have a genetic defect called alpha$_1$ antitrypsine deficiency. COPD includes two major disorders: chronic bronchitis and emphysema. In most patients with COPD both disorders coexist.

Chronic bronchitis is characterized by cough and sputum which are present mainly in winter over at least 2 consecutive years. In some patients cough and sputum are the only symptoms, whereas others also complain of difficulty in breathing/shortness of breath. If you have a chronic cough and produce sputum ask your doctor to check your lungs.

Emphysema is an abnormality of the airway sacs/alveoli. The tissues surrounding the sacs and the walls of the sacs are affected and become distended and look like holes in the lung (Swiss cheese appearance). Shortness of breath is the main symptom. Cough can be present, but to a lesser extent than in chronic bronchitis, and the chest becomes distended (barrel shape).

Bronchospasm

Open airway

Mucus

Thickened bronchial wall

Normal alveoli

Damaged alveoli

Enlargement of lung tissue:
(*left*) From a normal lung.
(*right*) From a lung with emphysema. In emphysema, lung tissue is destroyed, resulting in fewer and larger alveoli.

Heart disease

Cigarette smoking is a major cause of coronary heart disease, which leads to heart attack. Smokers have a greater risk of developing chronic disorders such as **atherosclerosis** (clogged arteries) and other damaging effects to the cardiovascular system. Smoking increases the risk of coronary heart disease by itself, and by acting with other risk factors to greatly increase risk. The nicotine and carbon monoxide in cigarette smoke prevent the effective release of oxygen into the bloodstream, damaging the cardiovascular system in many ways.

Damage to the heart
Heart disease leading to heart attack:
Fat deposits and blockages form in the arteries that supply the heart with blood.
Congestive heart failure:
Heart becomes damaged and enlarged from working so hard to pump blood against the higher blood pressure.

Blocked coronary artery

Enlarged heart

Damaged heart tissue

Normal heart

Women's health issues:
Risk factors and pregnancy

Using tobacco increases a woman's risk of chronic health problems including pulmonary complications and premature death. Studies suggest that cigarette smoking dramatically increases the risk of heart disease among premenopausal women who are also taking birth control pills. Studies show that mothers who smoke a pack or more of cigarettes a day consistently produce smaller babies than do non-smokers. The carbon monoxide inhaled with cigarette smoke reaches the fetus and diminishes its ability to absorb oxygen, resulting in significant oxygen deprivation. Other complications include decreased blood flow, which diminishes the transfer of essential nutrients from mother to fetus. A small baby is generally weaker and more vulnerable to illness than one of average size. Smokers are more apt to have their pregnancy end in premature birth, miscarriage or stillbirth. Research also suggests that infants are more likely to die from **Sudden Infant Death Syndrome (SIDS)** if their mothers smoke during and after pregnancy.

Enlargement of placental tissue:
Nicotine stimulates the release of hormones that constrict the vessels supplying blood to he placenta and uterus, diminishing the transfer of essential nutrients from mother to fetus.

Cancers

Smoking exposes the body to many cancer-causing chemicals that flow through the body. Tobacco byproducts have been found in the cervical mucus in women who smoke. Researchers believe these substances damage the structure of the cells in the cervix and may contribute to the development of cancers.

Endometrium (lining of uterus)
Uterus
Cervical cancer
Cervix
Vagina

©Scientific Publishing Ltd., Elk Grove Village, IL USA
#1352

PLATE 15

Understanding Rhinitis

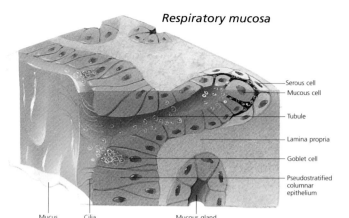

Respiratory mucosa

- Serous cell
- Mucous cell
- Tubule
- Lamina propria
- Goblet cell
- Pseudostratified columnar epithelium

Mucus — Cilia — Mucous gland

What is rhinitis?

Rhinitis is the inflammation of the mucous membranes that line the nose. The name rhinitis stems from the Greek word rhinos meaning "of the nose." This inflammation increases the production of nasal mucus and typically makes breathing through the nose difficult. Rhinitis can result from infections, such as the common cold and viral infections, allergic reactions, and unknown causes. Other cases of rhinitis result from common outdoor allergens such as airborne tree, grass or weed pollens. This is known as "hay fever." Certain *chronic* (long lasting) forms of rhinitis may cause the mucous membranes to thicken or to wear away.

Cross-section of paranasal sinuses

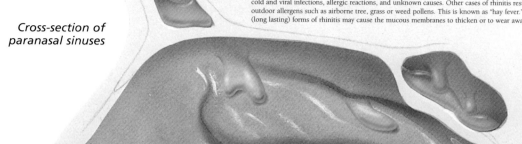

Inflamed respiratory mucosa

Inflammation

As we breathe, foreign particles, viruses and bacteria attach to the mucous membranes of sinuses to produce a clear mucus that washes the inside of the nose. In the event of a massive invasion of foreign micro-organisms, such as a virus, our body responds with a full defense called **inflammation**, characterized by heat, swelling and pain. The delivery of increased blood supply and heat from the interior of the body (fever) helps the body's defense system to function more effectively and slows the growth of the invading organisms. Fluid from the blood vessels leaks into the inflamed area, causing swelling and delivering antibodies. Invaders are broken down and devoured by granulocytes.

Allergic rhinitis
Immediate reaction *(in minutes)*

- Antigen attaches to antibodies
- Mast cell releases mediators

Immediate reaction
(in minutes)
After the immediate allergic reaction, the chemicals released by the mast cells also may produce a late (delayed) reaction in the airways. Inflammatory cells are attracted to the area and cause the airways to again narrow and swell. This delayed reaction can last much longer than the initial reaction.

Paranasal sinuses

- Frontal sinus
- Nasal cavity
- Anterior
- Middle — Ethmoid cells
- Posterior
- Opening of left maxillary sinus
- Sphenoid sinus
- Middle nasal meatus
- Inferior nasal meatus

Close-up of paranasal sinuses

- Antigen
- Mucus
- Cilia
- Antibodies attached to mast cell
- Epithelial cells
- Goblet cells

Taking control of your rhinitis
- Have a physician conduct a complete physical prior to diagnosis.
- Understand rhinitis, causes, symptoms and available treatments and medications.
- Take prescribed medications according to your plan.
- Identify and avoid triggers as much as possible.
- Be alert to possible symptoms.
- Always follow your physician's instructions.

- Frontal sinus
- Ethmoid cells
- Superior nasal concha
- Superior nasal meatus
- Middle nasal concha
- Middle nasal meatus
- Inferior nasal concha
- Inferior nasal meatus
- Nasal septum
- Maxillary sinus

Treatment of rhinitis

The ultimate goal in treating allergic and nonallergic rhinitis is to provide relief from the symptoms and prevent acute attacks from taking place. Most cases of mild allergic rhinitis require little more than reducing exposure to allergens and using a nasal wash. Fortunately there are a number of medications available, which can alleviate symptoms, with minimal side effects, for the more severe or chronic forms of rhinitis.

Treatments for rhinitis, as determined by your physician and based on your condition, include:

• Inhaled medications*
• Oral medications
• Immunotherapy (allergy shots)
• Surgery for some conditions

Relievers, also known as bronchodilators, are usually delivered by inhalers. They open constricted airways and provide temporary relief for symptoms, but don't relieve inflammation. They typically relieve symptoms for six hours, but some provide relief for up to twelve hours. *Examples of these medications include decongestant/antihistamine combinations.*

Preventors (anti-inflammatory medications) reduce inflammation in the airways and are taken continuously to help prevent or decrease the frequency of attacks. *Examples of these drugs are antihistamine tablets, nasal corticosteroids (commonly called steroids and considered to be the most effective in preventing allergy attacks) and nasal cromolyn for mild cases.*

Types of rhinitis:

Rhinitis is typically classified into the following categories:

Acute r. If symptoms last less than six weeks, this condition is described as acute rhinitis and typically results from a cold or other infection, or temporary exposure to environmental chemicals or pollutants.

Allergic r. is caused by "allergens" which are the triggers for allergies. Allergens can be found indoors or outside. Allergic **rhinitis** is caused by outdoor allergens. These symptoms are referred to as seasonal allergies.

Chronic r. If the condition lasts for a longer period, the condition is called chronic rhinitis. Chronic rhinitis often stems from indoor and outdoor allergies, but can also be caused by structural problems in the nasal septum or chronic infections.

Scrofulous r. A tuberculous infection of the nasal mucous membrane.

Vasomotor r. is a form of chronic nonallergic rhinitis and occurs when the nasal membrane swells in response to irritants such as smoke, environmental toxins or stress.

©Scientific Publishing Ltd., Elk Grove Village, IL USA
#1353

PLATE 16

The Common Cold vs. the Flu

What is a common cold?

The common cold is an infection of the nose and throat caused by a virus (a microscopic infectious agent or 'germ'). Colds can involve the **sinuses**, **ears**, **bronchial tubes** and the **eustachian tube**, which connects the middle ear and throat, windpipe, voice box and airways. Infection with a cold virus is less severe than influenza, but mild cases of flu can sometimes appear similar to a common cold. Cold symptoms include sneezing, runny or blocked nose, sore throat, cough and low grade fever. Most common colds occur gradually and last two to three days; severe colds can last up to two weeks.

How a cold virus infection occurs

The cold virus is present in the mucus and saliva of an infected person. You can catch a cold when the cold virus enters your nasal passages. This may be through **contaminated fingers** or **airborne droplets from coughs and sneezes**. The virus enters healthy cells in the nose and the adenoid, which become infected and produce new virus particles. As cells die, new cold viruses are released and infect other cells. Only small doses of the virus (2-30) particles are necessary to produce an infection. At present, over 200 cold viruses have been identified.

Treating a cold

There are no medications currently available that can cure the common cold, but symptoms can be treated or alleviated with nonprescription **analgesics** such as paracetamol and ibuprofen, as well as over-the-counter **decongestants**, **cough syrups** and **throat lozenges**. (Check with your physician prior to taking any medication if you are already taking prescription or over-the-counter drugs.) Plenty of rest and fluids to thin mucus secretions are also important.

Aspirin and colds/flu

Aspirin should never be given to children or teenagers who have cold or flu-like symptoms, particularly fever, due to the risk of a rare but dangerous complication of influenza known as **Reyes syndrome**.

Inflammation of bronchial airway

- Bronchospasm
- Mucus
- Open airway
- Thickened bronchial wall

What is the flu?

"Flu" is the common name for **influenza**, a viral illness of the **respiratory tract**. Infection occurs in the **nose**, **throat**, **bronchial tubes** and **lungs**. Symptoms, such as **muscle aches**, **fever** and **chills**, usually begin suddenly, are more severe and last longer than common cold symptoms. Because infection with an influenza virus can damage the lungs, it can lead to significant complications such as **pneumonia**. Those most at risk of serious complications of influenza are children, older people (over 65), people with chronic illnesses (such as asthma, heart disease or diabetes) and people with compromised immune systems.

How flu virus is transmitted

The **coughing**, **sneezing** and **talking** of an infected person releases millions of tiny droplets containing the virus into the air. Once inhaled, they enter the body and reproduce in healthy cells, infecting the lining of the respiratory tract. Influenza virus is **highly contagious** and can be transmitted up to seven days after symptoms begin.

Respiratory mucosa

- Serous cell
- Mucous cell
- Tubule
- Lamina propria
- Goblet cell
- Pseudostratified columnar epithelium
- Mucus
- Cilia
- Mucous gland

Inflamed respiratory mucosa

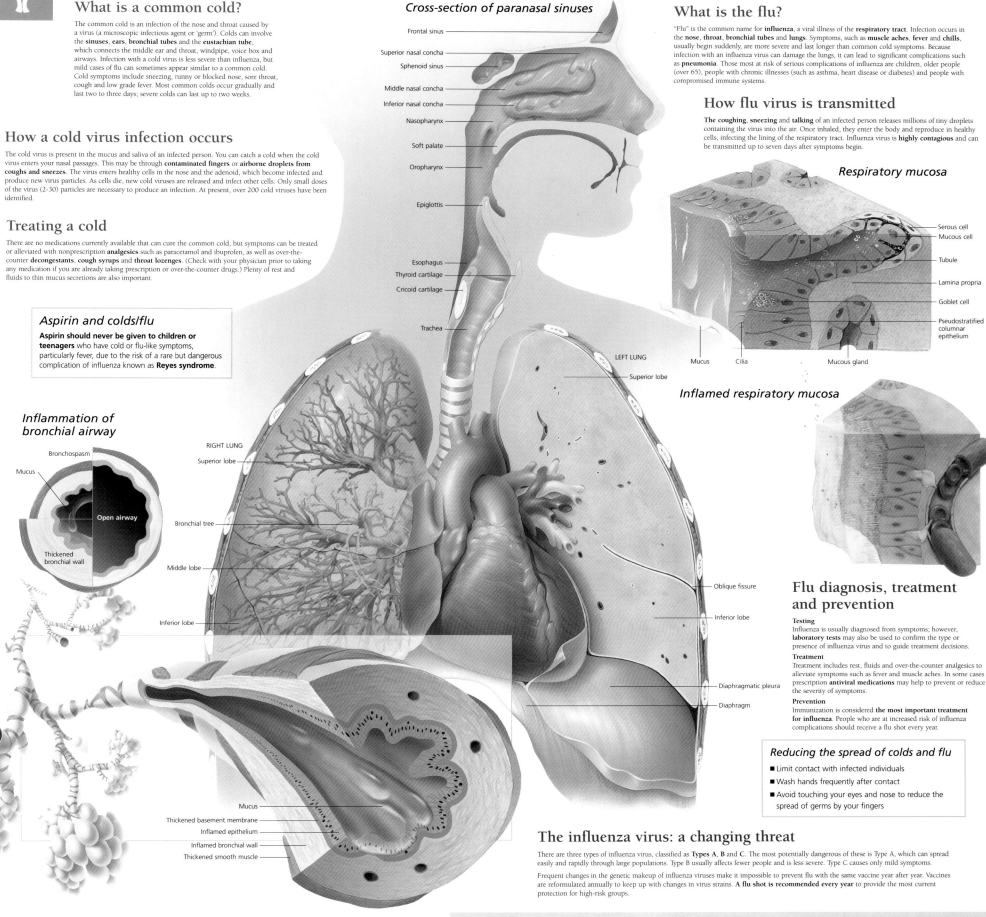

Cross-section of paranasal sinuses

- Frontal sinus
- Superior nasal concha
- Sphenoid sinus
- Middle nasal concha
- Inferior nasal concha
- Nasopharynx
- Soft palate
- Oropharynx
- Epiglottis
- Esophagus
- Thyroid cartilage
- Cricoid cartilage
- Trachea
- LEFT LUNG
  - Superior lobe
- RIGHT LUNG
  - Superior lobe
- Bronchial tree
- Middle lobe
- Inferior lobe
- Oblique fissure
- Inferior lobe
- Diaphragmatic pleura
- Diaphragm
- Mucus
- Thickened basement membrane
- Inflamed epithelium
- Inflamed bronchial wall
- Thickened smooth muscle

Flu diagnosis, treatment and prevention

Testing
Influenza is usually diagnosed from symptoms; however, **laboratory tests** may also be used to confirm the type or presence of influenza virus and to guide treatment decisions.

Treatment
Treatment includes rest, fluids and over-the-counter analgesics to alleviate symptoms such as fever and muscle aches. In some cases prescription **antiviral medications** may help to prevent or reduce the severity of symptoms.

Prevention
Immunization is considered **the most important treatment for influenza**. People who are at increased risk of influenza complications should receive a flu shot every year.

Reducing the spread of colds and flu

- Limit contact with infected individuals
- Wash hands frequently after contact
- Avoid touching your eyes and nose to reduce the spread of germs by your fingers

The influenza virus: a changing threat

There are three types of influenza virus, classified as **Types A**, **B** and **C**. The most potentially dangerous of these is Type A, which can spread easily and rapidly through large populations. Type B usually affects fewer people and is less severe. Type C causes only mild symptoms.

Frequent changes in the genetic makeup of influenza viruses make it impossible to prevent flu with the same vaccine year after year. Vaccines are reformulated annually to keep up with changes in virus strains. **A flu shot is recommended every year** to provide the most current protection for high-risk groups.

Middle ear infections

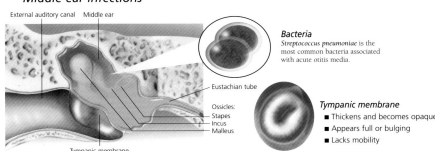

- External auditory canal
- Middle ear
- Eustachian tube
- Ossicles:
  - Stapes
  - Incus
  - Malleus
- Tympanic membrane

Bacteria
Streptococcus pneumoniae is the most common bacteria associated with acute otitis media.

Tympanic membrane
- Thickens and becomes opaque
- Appears full or bulging
- Lacks mobility

Common cold vs. flu: *Understanding the difference*

COMMON COLD	INFLUENZA
Primary symptoms – gradual onset	**Primary symptoms – acute onset**
▪ Runny nose, sneezing, watery eyes	▪ Severe headache and/or muscle aches
▪ Sore or scratchy throat	▪ Fever of 39-40°C or 102-104°F (may last 3-4 days)
▪ Mild cough	▪ Fatigue/exhaustion (may last two weeks or more)
▪ Fatigue	▪ Chills
▪ Occasional low grade fever (<38.5°C or 101°F)	▪ Cough (can become severe)
Complications	*Complications*
▪ Middle ear infections (otitis media)	▪ Pneumonia
▪ Acute bacterial sinusitis	▪ Bronchial, ear and sinus infections
▪ Asthma attacks and worsening of chronic bronchitis	▪ Worsening of chronic conditions such as asthma, congestive heart failure and diabetes
Prevention	*Prevention*
▪ No preventive treatment	▪ Annual flu shots recommended for high-risk groups
	▪ Sometimes antiviral medications may be used to prevent flu and reduce symptom duration and severity

©Scientific Publishing Ltd., Elk Grove Village, IL. USA
#1356

PLATE 17

The Heart

The heart is a four-chambered, muscular organ that functions as a powerful pump. About the size of a fist, the heart is located in the chest between the lungs, just to the left of center. The heart continuously pumps blood through the body's extensive network of arteries and veins. Arteries transport blood away from the heart, and veins transport blood back to the heart. This circulation of blood delivers oxygen and nutrients to the body while removing waste products.

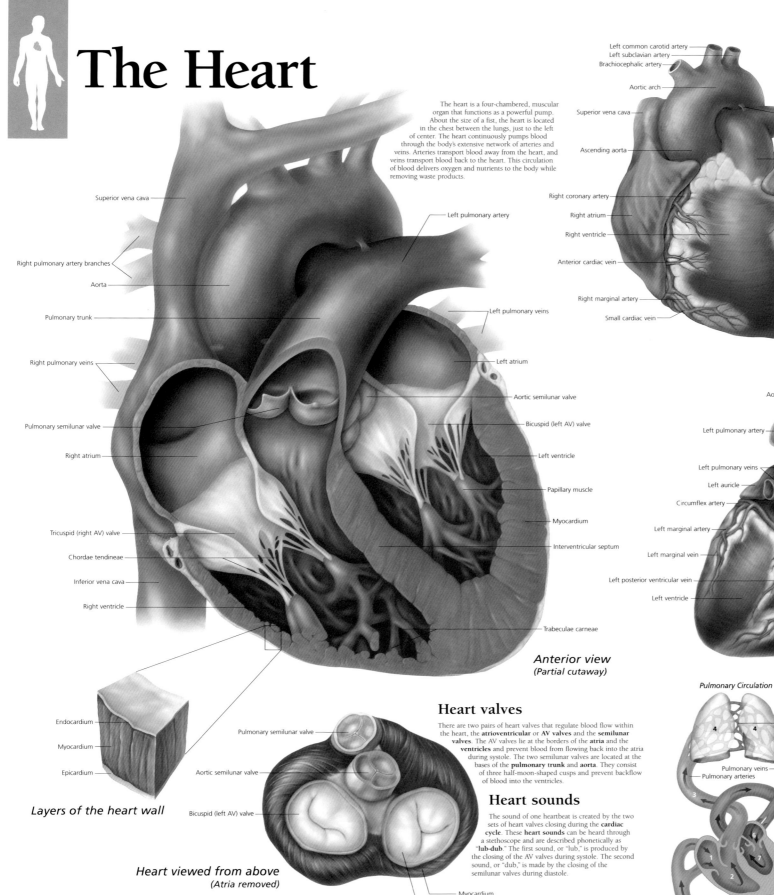

Anterior view (Partial cutaway)

- Superior vena cava
- Right pulmonary artery branches
- Aorta
- Pulmonary trunk
- Right pulmonary veins
- Pulmonary semilunar valve
- Right atrium
- Tricuspid (right AV) valve
- Chordae tendineae
- Inferior vena cava
- Right ventricle
- Left pulmonary artery
- Left pulmonary veins
- Left atrium
- Aortic semilunar valve
- Bicuspid (left AV) valve
- Left ventricle
- Papillary muscle
- Myocardium
- Interventricular septum
- Trabeculae carneae

Blood supply to the heart

The **coronary arteries** supply the **myocardium**, the muscular part of the heart wall, with oxygen and nutrients. These arteries originate from the aorta and lie within the **epicardium**, the outermost layer of the heart wall. Smaller arterial branches penetrate the myocardium. The **cardiac veins** collect venous blood from the heart wall and return it to the right atrium.

Anterior view

- Left common carotid artery
- Left subclavian artery
- Brachiocephalic artery
- Aortic arch
- Superior vena cava
- Ascending aorta
- Right coronary artery
- Right atrium
- Right ventricle
- Anterior cardiac vein
- Right marginal artery
- Small cardiac vein
- Ligamentum arteriosum
- Left pulmonary artery
- Pulmonary trunk
- Left auricle
- Circumflex artery
- Great cardiac vein
- Anterior descending (interventricular) artery
- Left ventricle
- Apex

Posterior view

- Aortic arch
- Left pulmonary artery
- Left pulmonary veins
- Left auricle
- Circumflex artery
- Left marginal artery
- Left marginal vein
- Left posterior ventricular vein
- Left ventricle
- Superior vena cava
- Right pulmonary artery
- Right pulmonary veins
- Right atrium
- Left atrium
- Inferior vena cava
- Coronary sinus
- Middle cardiac vein
- Posterior descending (interventricular) artery

Layers of the heart wall

- Endocardium
- Myocardium
- Epicardium

Heart valves

There are two pairs of heart valves that regulate blood flow within the heart, the **atrioventricular** or **AV valves** and the **semilunar valves**. The AV valves lie at the borders of the **atria** and the **ventricles** and prevent blood from flowing back into the atria during systole. The two semilunar valves are located at the bases of the **pulmonary trunk** and **aorta**. They consist of three half-moon-shaped cusps and prevent backflow of blood into the ventricles.

Heart viewed from above (Atria removed)

- Pulmonary semilunar valve
- Aortic semilunar valve
- Bicuspid (left AV) valve
- Myocardium
- Tricuspid (right AV) valve

Heart sounds

The sound of one heartbeat is created by the two sets of heart valves closing during the **cardiac cycle**. These **heart sounds** can be heard through a stethoscope and are described phonetically as "**lub-dub**." The first sound, or "lub," is produced by the closing of the AV valves during systole. The second sound, or "dub," is made by the closing of the semilunar valves during diastole.

Blood circulation

The heart functions as two side-by-side coordinated pumps. The right side of the heart receives carbon dioxide-rich blood from the body and pumps it to the lungs to be oxygenated. The left side of the heart receives the newly oxygenated blood from the lungs and pumps it throughout the entire body, delivering oxygen and nutrients to cells and tissues.

Pulmonary Circulation

- Pulmonary capillaries
- Pulmonary veins
- Pulmonary arteries
- Systemic capillaries

Systemic Circulation

- Carbon dioxide-rich blood
- Oxygen-rich blood

Progression of blood flow

1. Blood moves from the right atrium to the right ventricle by atrial contraction
2. Blood moves from the right ventricle to the pulmonary artery by ventricular contraction
3. Blood moves from the heart into the pulmonary artery towards the lungs
4. Blood moves into the capillaries in the lungs
5. Blood moves into the pulmonary vein back towards the heart
6. Blood moves from the left atrium to the left ventricle by atrial contraction
7. Blood moves from the left ventricle into the aorta by ventricular contraction
8. Blood moves into the capillaries in the body
9. Blood returns to the heart through the vena cavae

Electrical pathways

The steady beating of the heart is regulated by electrical impulses traveling through the heart. The impulses originate in the **sinoatrial node**, also known as the body's pacemaker. The impulses spread across the atria, causing them to contract. Next the impulses travel to the **atrioventricular node**, pause, then spread through the ventricles along special conduction pathways called **bundle branches** and **Purkinje fibers**. This causes the ventricles to contract.

Electrocardiogram

An **electrocardiogram** (**ECG** or **EKG**) graphically records the electrical activity of the heart. A typical ECG records three waves, each representing different phases in the **cardiac cycle**.

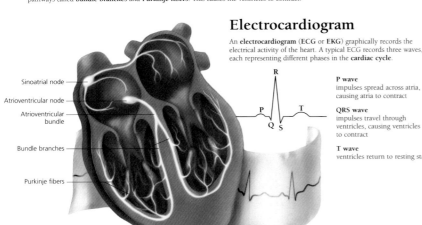

- Sinoatrial node
- Atrioventricular node
- Atrioventricular bundle
- Bundle branches
- Purkinje fibers

P wave
impulses spread across atria, causing atria to contract

QRS wave
impulses travel through ventricles, causing ventricles to contract

T wave
ventricles return to resting state

Cardiac cycle

The cardiac cycle refers to the alternating contraction and relaxation of the heart during one heartbeat. One cardiac cycle takes about four-fifths of a second to complete and repeats continuously. At rest, the heart beats an average of 60–80 times per minute. The cardiac cycle consists of two phases, **diastole** and **systole**. In diastole, the ventricles relax and fill with blood. In systole, the ventricles contract, forcing blood into the arteries.

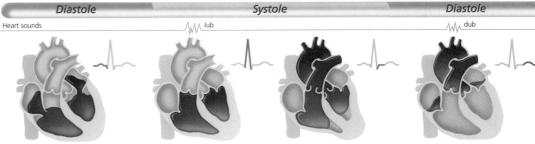

Diastole	Systole	Diastole
Heart sounds	lub	dub

Blood returns to the heart and flows into the atria. The pressure of the blood forces the AV valves open, and blood flows into the ventricles. The atria contract, forcing additional blood into the ventricles.

The atria relax and the ventricles begin to contract. Pressure rises in the ventricles, closing the AV valves. This closure causes the first heart sound.

Pressure continues to rise in the ventricles until it exceeds pressure in the arteries. Blood is forced out through the semilunar valves into the aorta and pulmonary trunk.

The ventricles relax, causing pressure in the ventricles to fall. Blood flowing back from the arteries closes the semilunar valves, causing the second heart sound. Blood begins to fill the atria again, and the cycle repeats.

©Scientific Publishing Ltd., Elk Grove Village, IL. USA
#1400

PLATE 18

High Blood Pressure

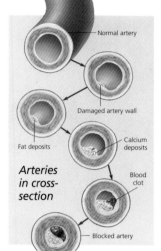

Capillaries · Venule · Waste

Tissue · Arteriole · Nutrients

Capillaries

In the venule, blood pressure is lower. Forces created by concentration differences cause waste products to return to the blood.

What is blood pressure?

Blood pressure is the force of circulating blood against the inner walls of the blood vessels. It is affected by:

- how hard the heart pumps
- the amount of blood in the body
- the diameter of the blood vessels

Generally, blood pressure increases when the heart pumps *harder,* the amount of blood in the body *increases* or the diameter of the blood vessels *decreases.*

Importance of pressure

Arterial blood carries essential materials such as oxygen and other nutrients to every cell in the body. Without an adequate supply of blood, organs and tissues cannot survive. Venous blood carries waste products away from the cells to be discarded. Both blood pressure and concentration must be within certain levels for this crucial exchange of nutrients and waste to occur. Fortunately, the body is armed with a web of complex mechanisms that monitor pressure and concentration and act to keep both within normal ranges.

Blood pressure forces blood into the tiny capillaries of the organs and contributes to the movement of nutrients out of the blood into the tissues.

What is high blood pressure?

One out of five adults in the U.S.—more than 50 million people—has high blood pressure. The term **hypertension** is also used to describe this condition, but it does not refer to being anxious or tense. It occurs when blood is flowing through the vessels at a pressure that is too high for the long-term health of the blood vessels. Generally, a blood pressure of 140 over 90 or higher is considered unhealthy. Over time, vessel walls exposed to these levels of pressure become damaged. This damage can lead to serious health problems.

What causes high blood pressure?

Occasionally, high blood pressure is caused by a disease. This type is called **secondary hypertension.** Most people with high blood pressure have a type called **essential** or **primary hypertension.** Although there are many theories about primary hypertension, the exact cause is unknown. It is possible that several complex mechanisms are involved.

Risk factors

Family history of high blood pressure

Race (African Americans have the highest incidence)

Age (risk increases with age)

Obesity

Sedentary lifestyle

Diabetes mellitus

Measuring blood pressure

Blood pressure is a measurement consisting of a top number, **systolic pressure** (pressure when the heart is contracting), and a bottom number, **diastolic pressure** (pressure when the heart is resting). It is measured with a pressure cuff and sphygmomanometer. The cuff is placed around the upper arm and tightened until blood flow through the brachial artery is stopped. Pressure, read from the attached meter, is gradually decreased in the cuff while a stethoscope is used to listen to the brachial artery. Sounds heard in the artery indicate the blood pressure. Blood pressure can be measured this way because blood makes noise when its flow is restricted:

Artery is blocked: no movement of blood—*silence*

Systolic pressure, artery begins to open: blood flow is turbulent—*sounds are heard*

Diastolic pressure, artery is completely open: blood flows smoothly through the artery—*silence*

The average blood pressure reading of a healthy adult is approximately:

$$\frac{120}{80}$$

Salt and blood pressure

Blood pressure and blood concentration of salt are closely related. When you eat salty foods, blood concentration goes up. Almost immediately, water is added to the blood by the kidneys so that blood concentration returns to normal. Additional water in the bloodstream elevates blood pressure. Blood pressure will stay elevated until the body is able to excrete the excess salt and water.

People with high blood pressure should watch their intake of salt. Although the exact role of salt is unknown, it is possible that some people with high blood pressure have a decreased ability to excrete it. Also, low salt intake may increase the effectiveness of medication.

Brachial artery · Pressure cuff · Stethoscope

Effects of high blood pressure

A person with high blood pressure usually has no symptoms until he or she has had it for quite some time and serious damage has occurred. For this reason, it is often called the "silent killer." Long-term damage from uncontrolled high blood pressure is often irreversible and can lead to an early death:

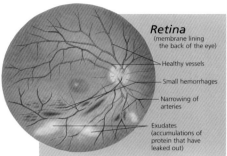

Retina (membrane lining the back of the eye)

- Healthy vessels
- Small hemorrhages
- Narrowing of arteries
- Exudates (accumulations of protein that have leaked out)

Damage to the eyes

It is unusual for eye damage from high blood pressure to impair vision, but the retina provides a remarkably accurate assessment of overall damage to blood vessels. The small blood vessels in the retina are a good sample of all the blood vessels in the body and are easily inspected.

Damage to the brain

- **Stroke:** A portion of brain tissue dies when it is deprived of blood supply. This can happen when a bulging artery (called an **aneurysm**) ruptures or an artery becomes blocked by a blood clot or fat deposits.
- **Cerebrovascular insufficiency:** A series of mini-strokes occurs in the smaller vessels of the brain. Tiny arterioles bulge, then burst from high pressure or become blocked by small blood clots. There are no symptoms until damage accumulates over time.

Damage to blood vessels

Artery walls become damaged from high pressure. Fat accumulates and the walls thicken. Calcium is deposited in the fatty areas, "hardening" the arteries, making them unable to increase in size. Blood flow through the arteries decreases. Damaged artery walls may also cause blood clots to form which block the artery itself or break off and block arteries in other organs.

Normal artery · Damaged artery wall · Fat deposits · Calcium deposits · Blood clot · Blocked artery

Arteries in cross-section

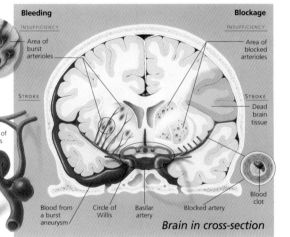

Bleeding · INSUFFICIENCY · Area of burst arterioles · STROKE

Blockage · INSUFFICIENCY · Area of blocked arterioles · STROKE · Dead brain tissue

Circle of Willis · Aneurysm · Basilar artery · Blood from a burst aneurysm · Circle of Willis · Basilar artery · Blocked artery · Blood clot

Brain in cross-section

Narrowing of renal artery

Diseased kidney

Blocked artery · Damaged heart tissue · Enlarged heart · Normal heart

Damage to the heart

- **Heart disease leading to heart attack:** Fat deposits and blockages form in the arteries that supply the heart with blood.
- **Congestive heart failure:** Heart becomes damaged and enlarged from working so hard to pump blood against the higher blood pressure.

Damage to the kidneys

- **Blood vessel damage:** Arteries become narrowed and stiff from high pressure. Blood flow to the kidneys is decreased. Receptors respond by recruiting mechanisms throughout the body to *raise* overall blood pressure even further.
- **Kidney disease leading to failure:** It becomes more and more difficult for the kidneys to remove impurities from the blood. Toxic materials accumulate.

Treatment of high blood pressure

The only way to detect high blood pressure early is to have your blood pressure measured by a healthcare professional. Secondary hypertension is treated by managing the disease that is causing it. Although primary hypertension cannot be cured, there are several ways to lower blood pressure and keep it controlled:

Diet and medication

Improvements in diet alone may be enough to control high blood pressure, especially in mild cases. Often, improvements in diet need to be combined with medication to control high blood pressure. Due to the vast number of mechanisms in the body that affect blood pressure, there are several different types of medications. Your physician will determine which is best for you and may eventually suggest taking more than one.

Diuretics decrease blood volume by causing more water and salt to be excreted in the urine.

Sympathetic nervous system blocking drugs cause the heart to slow down and beat less forcefully. They also decrease constriction of arteries throughout the body. *Examples are alpha, beta and adrenergic blockers.*

Vasodilators act directly on blood vessel walls or through other mechanisms to increase blood vessel diameter. *Examples are calcium channel blockers, ACE inhibitors, angiotensin II inhibitors, and angiotensin receptor blockers.*

Taking control of your blood pressure

- Maintain a low-fat diet.
- Decrease salt intake to less than a teaspoon per day (2000 mg).
- Shed extra weight to decrease strain on your heart.
- Don't smoke.
- Restrict caffeine and alcohol consumption.
- Follow all of your physician's instructions.
- Take prescribed medications as part of your daily routine.
- Consult your physician about an appropriate exercise plan and follow it.
- Measure blood pressure regularly at home.
- Continue taking medication even after your blood pressure has reached a good level.

Effective control of high blood pressure can prevent most of its complications.

©Scientific Publishing Ltd., Elk Grove Village, IL, USA
#1450

PLATE 19

Effects of Hypertension

Damage to the heart

- **Heart disease leading to heart attack:**
  Fat deposits and blockages form in the arteries that supply the heart with blood.

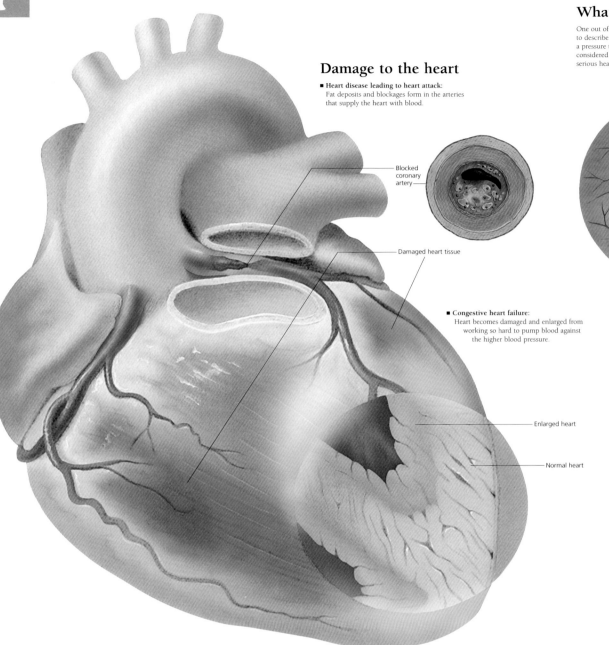

Blocked coronary artery

Damaged heart tissue

- **Congestive heart failure:**
  Heart becomes damaged and enlarged from working so hard to pump blood against the higher blood pressure.

Enlarged heart

Normal heart

What is high blood pressure?

One out of five adults in the U.S.—more than 50 million people—has high blood pressure. The term **hypertension** is also used to describe this condition, but it does not refer to being anxious or tense. It occurs when blood is flowing through the vessels at a pressure that is too high for the long-term health of the blood vessels. Generally, a blood pressure of 140 over 90 or higher is considered unhealthy. Over time, vessel walls exposed to these levels of pressure become damaged. This damage can lead to serious health problems.

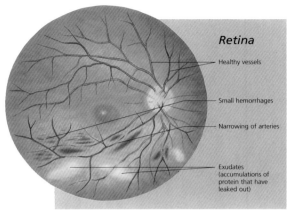

Retina

Healthy vessels

Small hemorrhages

Narrowing of arteries

Exudates (accumulations of protein that have leaked out)

Damage to the eyes

It is unusual for eye damage from high blood pressure to impair vision, but the retina provides a remarkably accurate assessment of overall damage to blood vessels. The small blood vessels in the retina are a good sample of all the blood vessels in the body and are easily inspected.

Damage to the kidneys

- **Blood vessel damage:**
  Arteries become narrowed and stiff from high pressure. Blood flow to the kidneys is decreased. Receptors respond by recruiting mechanisms throughout the body to *raise* overall blood pressure even further.

- **Kidney disease leading to failure:**
  It becomes more and more difficult for the kidneys to remove impurities from the blood. Toxic materials accumulate.

Narrowing of renal artery

Diseased kidney

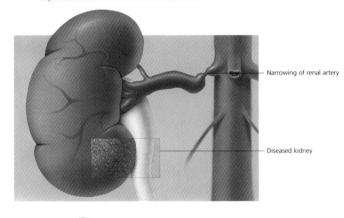

Capillaries

Venule

Tissue

Arteriole

Capillaries

Damage to the brain

- **Stroke:**
  A portion of brain tissue dies when it is deprived of blood supply. This can happen when a bulging artery (called an **aneurysm**) ruptures or an artery becomes blocked by a blood clot or fat deposits.

- **Cerebrovascular insufficiency:**
  A series of mini-strokes occurs in the smaller vessels of the brain. Tiny arterioles bulge, then burst from high pressure or become blocked by small blood clots. There are no symptoms until damage accumulates over time.

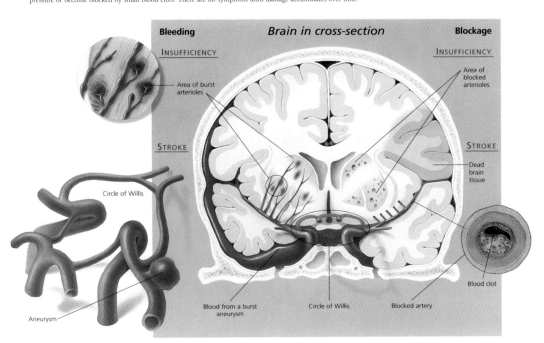

Bleeding	*Brain in cross-section*	Blockage
INSUFFICIENCY		INSUFFICIENCY

Area of burst arterioles

Area of blocked arterioles

STROKE

STROKE

Dead brain tissue

Circle of Willis

Aneurysm

Blood from a burst aneurysm

Circle of Willis

Blocked artery

Blood clot

Damage to blood vessels

Artery walls become damaged from high pressure. Fat accumulates and the walls thicken. Calcium is deposited in the fatty areas, "hardening" the arteries, making them unable to increase in size. Blood flow through the arteries decreases. Damaged artery walls may also cause blood clots to form which block the artery itself or break off and block arteries in other organs.

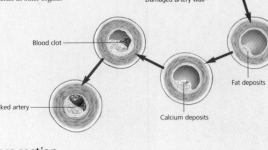

Normal artery

Damaged artery wall

Blood clot

Fat deposits

Blocked artery

Calcium deposits

Arteries in cross-section

©Scientific Publishing Ltd., Elk Grove Village, IL, USA
#1451

PLATE 20

Understanding Angina

What is angina?

Angina is chest pain due to restricted blood flow to the **myocardium** (heart muscle). Restricted blood flow may come from blockages in the arteries that supply blood to the heart. This condition is called **coronary artery disease**. Angina usually occurs during physical activities, eating or stress and is often accompanied by sweating and difficulty in breathing. Angina is a symptom, not a disorder. Not all chest pain is angina; your physician can make the proper diagnosis.

Types of angina:

Stable angina -
The most common variety of angina that follows somewhat predictable patterns.

Unstable angina -
Prolonged angina with symptoms that suggests a heart attack without physical or diagnostic evidence. Less responsive to rest and medication.

Prinzmetal's angina -
A rare form of angina that is caused by vasospasm, a spasm that narrows the coronary artery to the heart.

Microvascular angina -
Poor functioning of tiny blood vessels in the heart that do not supply proper blood flow. Not associated with typical coronary artery blockages.

Blocked artery

Angina warning symptoms include:

- Pain that seems to affect your entire chest area.
- Pain that seems to randomly move through the arm, neck and shoulders and possibly other parts of the body.
- A feeling of nausea, indigestion or heartburn.
- Pain lasting no more than 15 minutes.
- Light-headed feeling, with a sense of anxiety and nervousness.

Damaged heart tissue

Narrowed artery

Ischemia
Oxygen-deprived heart muscle caused from narrowed artery

A coronary angiogram of a heart. The circle indicates a narrowing of a major artery due to atherosclerosis.

Healthy heart
(Anterior view)

Left common carotid artery
Brachiocephalic artery
Left subclavian artery
Aortic arch
Superior vena cava
Left pulmonary artery
Ascending aorta
Pulmonary trunk
Left auricle
Right coronary artery
Circumflex artery
Right atrium
Great cardiac vein
Right ventricle
Anterior descending (interventricular) artery
Anterior cardiac vein
Left ventricle
Right marginal artery
Small cardiac vein
Apex

Blood supply to the heart

The **coronary arteries** supply the **myocardium**, the muscular layer of the heart wall, with oxygen and nutrients. These arteries originate from the aorta and lie within the **epicardium**, the outermost layer of the heart wall. Smaller arterial branches penetrate the myocardium. The **cardiac veins** collect venous blood from the heart wall and return it to the right atrium.

What causes angina?

Angina is caused by insufficient blood flow to the heart. This could be the result of an artery obstruction due to a buildup of plaque, known as atherosclerosis. It can be the result of diabetes, an inactive lifestyle or smoking. Physical activity may provoke the attack in a person with an existing risk for angina. Other less common causes of angina are abnormal heart rhythms or heart valves.

Damage to the blood vessels *(Atherosclerosis)*

Stable plaque

One of the primary causes of angina is atherosclerosis, which most people term "hardening of the arteries." This is a buildup of cholesterol, minerals, blood and muscle cells in the artery wall. The presence of these substances causes the artery to become stiffer and can cause the blood flow to be blocked to some degree. Plaque is the term used to describe this buildup. A minimal amount of plaque may not cause significant harm; this stage is known as stable plaque.

Ruptured plaque

If plaque develops further and breaks up, termed ruptured plaque, it may have quick and damaging effects. When the rupture occurs, it releases harmful enzymes and other substances, potentially causing a blood clot. The blood clot may restrict the artery at the point of the rupture or travel through the bloodstream to cause a stroke, a heart attack or kidney failure.

100% blockage

Plaque formation can progress if unchecked to completely block the artery. This may prevent needed oxygen from reaching vital heart muscle tissue. The lack of oxygen may lead to debilitating heart conditions.

Damage to the heart

Heart disease leading to heart attack: Fat deposits and blockages form in the arteries that supply the heart with blood.
Congestive heart failure: Heart becomes damaged and enlarged from working so hard to pump blood against the higher blood pressure.

Diagnosing angina

The initial warning sign is chest pain during physical activity. However, anytime a feeling of chest tightness occurs, whether during mild exertion or even in sleep, angina could be the cause. It is sometimes confused with indigestion, but commonly causes pain in shoulders, neck and arms.

The most accurate way to assess angina is by a **coronary angiogram**, an x-ray of the coronary artery. A dye is injected into the bloodstream and the x-ray shows the coronary arteries and their narrowing.

Physicians might use an **electrocardiogram** or ECG to evaluate the heart for damage. The next procedure after an electrocardiogram may be a **stress test**. The ECG is performed while walking on a treadmill.

There are also heart enzyme levels the physician can examine to eliminate a diagnosis of angina. Other conditions such as anemia or thyroid abnormalities might place stress on the heart and cause angina.

The treatment of angina

Lifestyle changes
A healthy lifestyle is a primary component in the treatment of angina. This includes eating habits and exercise. Those who are active and control their weight lessen their risk of angina.

Physicians may prescribe cardiac medications such as nitroglycerin to increase the blood flow to the coronary system by expanding coronary blood vessels. Beta-adrenergic blockers might be used to slow the heartbeat, relieving stress on the heart. Physicians might suggest aspirin because it is known to dissolve blood clots.

Cholesterol control
The control of "bad" cholesterol or LDL may be instrumental in preventing future attacks. Physicians most often prescribe statins to reduce LDL to 100 mg/dL or below. They have limited side effects, but must be used with a low-fat diet to achieve the best results.

Smokers may reduce their incidence of heart attack by 50% if they are smoke-free for at least a year. The longer the period without smoking the greater the benefit. After five to ten years, their risk of a heart attack is the same as for someone who has never smoked.

Taking control of your angina

- Controlling physical activity – *Consult your physician about an appropriate exercise plan and follow it.*
- Shed extra weight to decrease strain on your heart.
- Maintain a low-fat diet.
- Decrease salt intake to less than a teaspoon per day.
- Don't smoke.
- Restrict caffeine and alcohol consumption.
- Follow all of your physician's instructions.
- Avoid any unnecessary emotional disturbances.
- Take prescribed medications as part of your daily routine.
- Measure blood pressure regularly at home.

Exercising has many benefits, especially reducing the risk of heart problems. Exercise may improve the levels of "good" or HDL cholesterol and reduce blood pressure in most cases. Patients should always consult their physician on the type of exercise program.

Surgical procedures might include a heart bypass, which requires placing veins in such a way to replace the function of the obstructed artery. A less extensive procedure is balloon angioplasty. This is when a catheter is inserted in an affected artery and a balloon is inflated to expand the passageway. A common second step in the procedure is to place a stent in a passage to give the artery the best chance of staying open. Additional surgical procedures utilize a catheter to enter the blocked artery and eliminate the obstruction with a laser (laser angioplasty) or revolving cleaning tool (atherectomy).

The diagram illustrates a restricted artery and the two-part procedure to improve blood flow. The artery shows a blockage (a) and the use of a balloon angioplasty (b) as well as the use of a stent (c).

Atherectomy is a surgical procedure that removes plaque buildup in arteries using a revolving cleaning tool.

Electrical pathways

The steady beating of the heart is regulated by electrical impulses traveling through the heart. The impulses originate in the **sinoatrial node**, also known as the body's pacemaker. The impulses spread across the atria, causing them to contract. Next the impulses travel to the **atrioventricular node**, pause, then spread through the ventricles along special conduction pathways called **bundle branches** and **Purkinje fibers**. This causes the ventricles to contract.

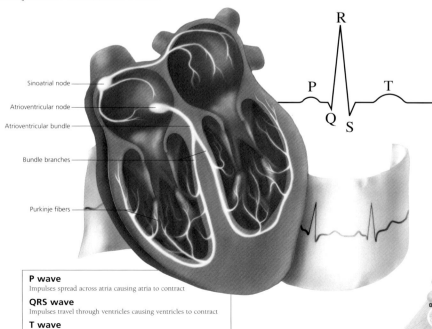

Sinoatrial node
Atrioventricular node
Atrioventricular bundle
Bundle branches
Purkinje fibers

R
P   T
Q   S

P wave
Impulses spread across atria causing atria to contract

QRS wave
Impulses travel through ventricles causing ventricles to contract

T wave
Ventricles return to resting state

Diet and medication

Improvements in diet and exercise alone may not be enough to control angina.

Often, improvements in diet need to be combined with controlling or lowering other risk factors, such as high blood pressure, cigarette smoking and high cholesterol levels.

Due to the vast number of mechanisms in the body that are associated with angina, there are several different types of medications. Your physician will determine which is best for you and may eventually suggest taking more than one.

Medications can significantly reduce both the effects and risk of angina attacks. These also decrease the chances that a patient with angina will have a heart attack.

Antiplatelets decrease the ability of the blood to clot. *Examples are aspirin, Heparin and IIb/IIIa.*

Sympathetic nervous system blocking drugs cause the heart to slow down and beat less forcefully. They also decrease constriction of arteries throughout the body. *Examples are alpha, beta and adrenergic blockers.*

Vasodilators act directly on blood vessel walls or through other mechanisms to increase blood vessel diameter. *Examples are calcium channel blockers and nitroglycerin.*

©Scientific Publishing Ltd., Elk Grove Village, IL, USA
#1453

PLATE 21

Understanding Heart Disease

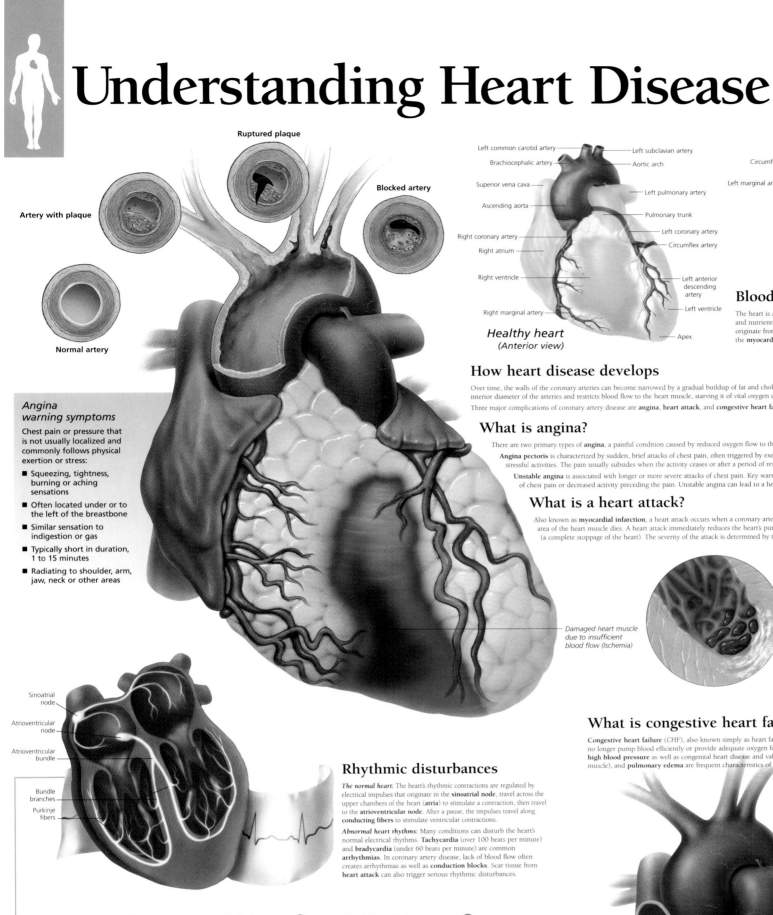

Ruptured plaque

Blocked artery

Artery with plaque

Normal artery

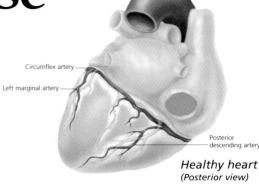

Circumflex artery

Left marginal artery

Posterior descending artery

Healthy heart
(Posterior view)

Left common carotid artery — Left subclavian artery
Brachiocephalic artery — Aortic arch
Superior vena cava
Ascending aorta — Left pulmonary artery
— Pulmonary trunk
Right coronary artery — Left coronary artery
Right atrium — Circumflex artery
Right ventricle — Left anterior descending artery
Right marginal artery — Left ventricle
— Apex

Healthy heart
(Anterior view)

Blood supply and the heart

The heart is a powerful muscle that depends on a continuous flow of oxygen and nutrients. This blood supply is provided by **coronary arteries**, which originate from the aorta and branch out to deliver oxygenated blood throughout the **myocardium**, the muscular layer of the heart wall.

How heart disease develops

Over time, the walls of the coronary arteries can become narrowed by a gradual buildup of fat and cholesterol deposits called **plaque**. This process, **atherosclerosis**, reduces the interior diameter of the arteries and restricts blood flow to the heart muscle, starving it of vital oxygen and nutrients. The resulting condition is **coronary artery disease (CAD)**. Three major complications of coronary artery disease are **angina**, **heart attack**, and **congestive heart failure**.

What is angina?

There are two primary types of **angina**, a painful condition caused by reduced oxygen flow to the muscle fibers in the heart.
Angina pectoris is characterized by sudden, brief attacks of chest pain, often triggered by exercise and other strenuous or stressful activities. The pain usually subsides when the activity ceases or after a period of rest.
Unstable angina is associated with longer or more severe attacks of chest pain. Key warning signs are changing patterns of chest pain or decreased activity preceding the pain. Unstable angina can lead to a heart attack.

What is a heart attack?

Also known as **myocardial infarction**, a heart attack occurs when a coronary artery is suddenly blocked by a blood clot. Deprived of oxygen, the affected area of the heart muscle dies. A heart attack immediately reduces the heart's pumping ability and may lead to **cardiac arrythmias** and **cardiac arrest** (a complete stoppage of the heart). The severity of the attack is determined by the amount of heart muscle damage and the function of nearby arteries.

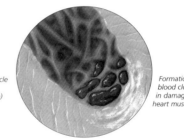

Damaged heart muscle due to insufficient blood flow (Ischemia)

Formation of blood clots in damaged heart muscle

Angina warning symptoms

Chest pain or pressure that is not usually localized and commonly follows physical exertion or stress:

- Squeezing, tightness, burning or aching sensations
- Often located under or to the left of the breastbone
- Similar sensation to indigestion or gas
- Typically short in duration, 1 to 15 minutes
- Radiating to shoulder, arm, jaw, neck or other areas

Heart attack warning symptoms

Pain, tightness, or sensations of fullness or squeezing in the chest that occur suddenly and last several minutes or more:

- Chest pain radiating to the jaw, shoulder or arm
- Shortness of breath
- Nausea and sweating
- Pale, clammy skin
- Weakness and light-headedness

What is congestive heart failure?

Congestive heart failure (CHF), also known simply as heart failure, is a condition in which the heart has become weak and can no longer pump blood efficiently or provide adequate oxygen for the brain and other organs. Causes of CHF include CAD and **high blood pressure** as well as congenital heart disease and valve disease. Enlarged heart, **hypertrophy** (thickening of the heart muscle), and **pulmonary edema** are frequent characteristics of CHF.

Rhythmic disturbances

Sinoatrial node
Atrioventricular node
Atrioventricular bundle
Bundle branches
Purkinje fibers

The normal heart: The heart's rhythmic contractions are regulated by electrical impulses that originate in the **sinoatrial node**, travel across the upper chambers of the heart (**atria**) to stimulate a contraction, then travel to the **atrioventricular node**. After a pause, the impulses travel along **conducting fibers** to stimulate ventricular contractions.

Abnormal heart rhythms: Many conditions can disturb the heart's normal electrical rhythms. **Tachycardia** (over 100 beats per minute) and **bradycardia** (under 60 beats per minute) are common **arrhythmias**. In coronary artery disease, lack of blood flow often creates arrhythmias as well as **conduction blocks**. Scar tissue from **heart attack** can also trigger serious rhythmic disturbances.

Atrial fibrillation

Ventricular fibrillation

Heart block fibrillation

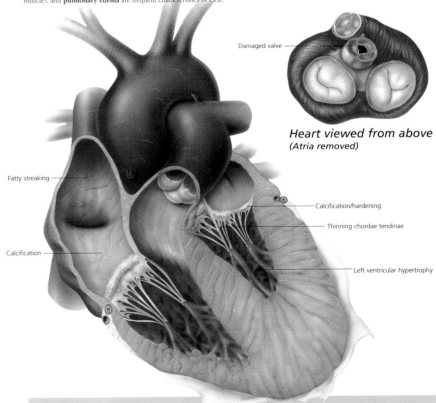

Damaged valve

Heart viewed from above
(Atria removed)

Fatty streaking

Calcification

Calcification/hardening

Thinning chordae tendinae

Left ventricular hypertrophy

Arrhythmia warning symptoms

Signs of a rhythmic disturbance may include palpitations, often described as fluttering in the chest (usually not serious unless multiple beats are skipped in succession). Other signs include pounding or racing heartbeat not related to exercise; repeated episodes of light-headedness or dizziness; sudden fainting spells; and cardiac collapse, a life-threatening condition.

Causes of coronary artery disease

The disease processes that lead to CAD have been attributed to many different causes. They include **genetic**, **age**, and **gender factors** as well as lifestyle or **modifiable risks**.

- High blood pressure
- Smoking
- High cholesterol
- Obesity
- Diabetes
- Inactive lifestyle
- Stress

A higher number of risk factors increases your chances of developing CAD.

Healthy lifestyle changes

By following recommended lifestyle changes, many risk factors for heart disease can be significantly reduced or controlled. Seeking treatment for **hypertension**, lowering **cholesterol** through diet and medication, quitting **smoking**, increasing **physical activity**, managing **diabetes**, and **losing weight** are all important steps to a healthier heart.

CHF warning symptoms

The most common signs of congestive heart failure include:

- Shortness of breath during regular activity (dyspnea)
- Frequent coughing
- Nausea and loss of appetite
- Increased heart rate
- Difficulty breathing while resting or sleeping
- Fluid buildup in feet, ankles, or legs (edema)
- Impaired thinking and confusion
- General fatigue

©Scientific Publishing Ltd., Elk Grove Village, IL, USA
#1454

PLATE 22

Understanding Stroke

What is stroke?

A **stroke** is a cerebral vascular accident that occurs when blood flow to the brain is suddenly interrupted by a **burst blood vessel** or a **blockage** in the brain's blood supply. Nerve cells in the affected part of the brain no longer receive oxygen and nutrients, and the result is temporary or permanent loss of function in the corresponding parts of the body.

Strokes are classified into two major categories:

• **Ischemic** stroke is the most common type, occurring in approximately 80 percent of all cases

• **Hemorrhagic** stroke is present in about 20 percent of stroke cases

What are the causes of stroke?

Every type of stroke has a specific physiological cause. In general, however, strokes are frequently caused by underlying medical conditions such as **high blood pressure**, **heart disease**, or **atherosclerosis** (narrowing of the arteries). Strokes may also be the result of **head injuries**, **aneurysms**, or **congenital defects** in the arteries of the brain.

HEMORRHAGIC

ISCHEMIC

Ischemic strokes happen when blood flow to the brain is blocked by clots or fragments that have become lodged within the blood vessels. The **origin of the clot** determines what type of ischemic stroke has occurred.

*Embolic strokes are frequently caused by **atrial fibrillation**, a heart condition in which incomplete pumping of the heart's upper chambers results in the formation of clots.*

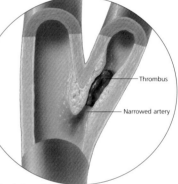

Embolus

Cerebral embolism

A blood clot called an **embolus** forms in the circulatory system. Blockage occurs when the clot reaches vessels in the brain too small to let it pass.

Headaches and seizures can occur almost immediately.

There are three types of **hemorrhagic stroke**, which occurs when blood from a ruptured vessel accumulates and compresses surrounding brain tissue, injuring cells and interfering with brain function. The leaking vessel also interrupts **oxygen flow** to the brain. The amount of bleeding determines the severity of the stroke.

Treating high blood pressure, which strains the blood vessels and increases the risk of stroke, is one of the most important ways to help prevent risk of stroke.

Normal   AVM

Thrombus

Narrowed artery

Cerebral thrombosis

A blood clot or **thrombus** forms within an artery supplying blood to the brain. The most common type of stroke, thrombosis often results from damage to the arteries caused by fatty deposit buildups (atherosclerosis).

Symptoms such as loss of feeling, speech problems and seizures may occur gradually, over a period of minutes or hours.

Circle of Willis

Burst aneurysm

Basilar artery

Subarachnoid

A ruptured blood vessel on the surface of the brain bleeds into the space **between the skull and the brain**.

As the vessel weakens, warning signs such as sudden headaches and light-sensitivity may be present for days or weeks.

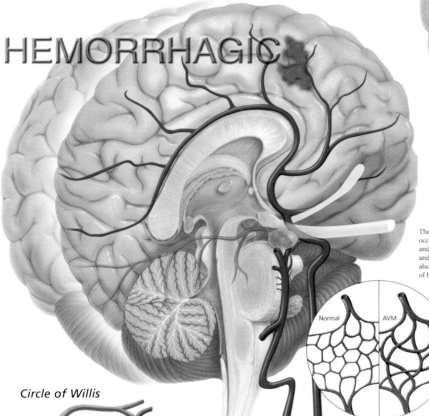

Area of burst arterioles

Brain in cross-section

Arteriovenous malformation (AVM)

A rupture occurs within a cluster of abnormally formed blood vessels in the brain.

Symptoms may include migraine-like headache, numbness, muscle weakness and seizures.

Intracerebral

Blood from a **ruptured artery in the brain** is released into surrounding brain tissue.

The symptoms usually occur suddenly and can include headache, nausea and marked changes in mental state.

Cerebrovascular insufficiency:

A series of mini-strokes occurs in the smaller vessels of the brain. Tiny arterioles bulge, then burst from high pressure or become blocked by small blood clots.

There are no symptoms until damage accumulates over time.

Stroke risk factors

■ High blood pressure
■ High blood cholesterol
■ Atherosclerosis
■ Heart disease/heart abnormalities
■ Adult-onset diabetes
■ Family history of stroke
■ Previous TIA
■ Cigarette smoking
■ Excess weight
■ Heavy alcohol consumption
■ Inactive lifestyle

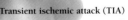

Transient ischemic attack (TIA)

A temporary blockage of blood flow to the brain is caused by **small emboli** that break up and dissolve shortly after lodging in vessel walls. Mild stroke-like symptoms can last for minutes or up to 24 hours.

Symptoms lasting over 24 hours are considered a stroke. TIAs are often warning signs of future ischemic stroke and should be treated immediately.

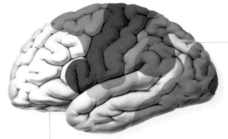

Functional areas of the brain

■ *Primary motor area*
■ *Secondary motor area*
■ *Primary somatosensory area*
■ *Secondary somatosensory area*
■ *Primary visual area*
■ *Secondary visual area*
■ *Primary acoustic area*
■ *Secondary acoustic area*
■ *Sensory speech area*

Effects of strokes

Three factors influence the effects of a stroke:

• What type of stroke was it?
• Where did the stroke occur?
• How much injury was caused?

The location of the stroke is especially critical in understanding which parts of the body will be most affected. A stroke near the back of the brain will often cause changes in vision. A stroke on the right side of the brain will affect neurological function on the left (opposing) side of the body.

Right brain strokes may cause:

• Paralysis of the left arm, leg and side of face
• Loss or impairment of analytical skills
• Problems with spatial perception
• Sudden, impulsive behavior
• Short-term memory loss

Left brain strokes may cause:

• Paralysis of the right arm, leg and side of face
• Difficulty speaking or understanding language (aphasia)
• Slow and cautious behavior
• Difficulty with conceptual thinking
• Memory loss and difficulty learning new tasks

Stroke symptoms

Most strokes share these warning signs and symptoms:

• Sudden or severe headache
• Dizziness or loss of balance
• Double vision or blurring in one or both eyes
• Difficulty in swallowing

• Weakness or numbness on one side of the body
• Difficulty in speaking or understanding others
• Confusion or difficulty thinking
• Sudden loss of bowel or bladder control

Call 911 immediately for help if you or someone you know experiences any of these symptoms of stroke.

Stroke rehabilitation

The time it takes to recover from a stroke and the amount of recovery possible depend on the amount of damage that occurred to the brain.

Regardless of the type of stroke, early intervention and treatment—within a few hours of the onset of the stroke–are critical in possibly preventing further brain injury and promoting long-term recovery.

• Diagnostic tests such as CT scans and MRI are used to determine the nature and extent of the stroke.

• Stroke rehabilitation can include physical, speech and occupational therapies.

• Recovery usually begins within the first few weeks and speech and function may continue to improve gradually for a year or more after the stroke.

©Scientific Publishing Ltd., Elk Grove Village, IL USA
#1455

PLATE 23

The Digestive System

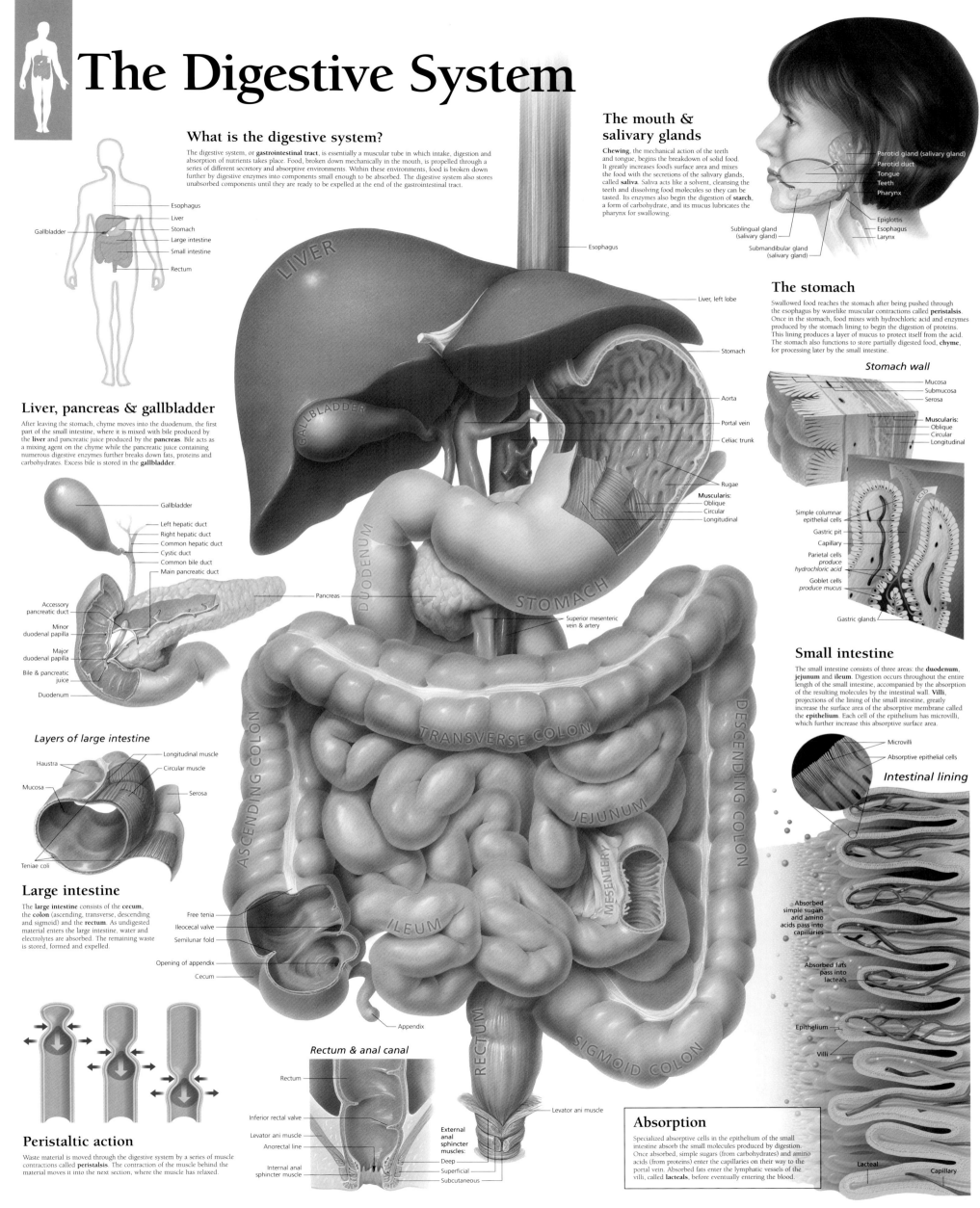

What is the digestive system?

The digestive system, or **gastrointestinal tract**, is essentially a muscular tube in which intake, digestion and absorption of nutrients takes place. Food, broken down mechanically in the mouth, is propelled through a series of different secretory and absorptive environments. Within these environments, food is broken down further by digestive enzymes into components small enough to be absorbed. The digestive system also stores unabsorbed components until they are ready to be expelled at the end of the gastrointestinal tract.

Esophagus
Liver
Gallbladder
Stomach
Large intestine
Small intestine
Rectum

The mouth & salivary glands

Chewing, the mechanical action of the teeth and tongue, begins the breakdown of solid food. It greatly increases food's surface area and mixes the food with the secretions of the salivary glands, called **saliva**. Saliva acts like a solvent, cleansing the teeth and dissolving food molecules so they can be tasted. Its enzymes also begin the digestion of **starch**, a form of carbohydrate, and its mucus lubricates the pharynx for swallowing.

Parotid gland (salivary gland)
Parotid duct
Tongue
Teeth
Pharynx
Sublingual gland (salivary gland)
Submandibular gland (salivary gland)
Epiglottis
Esophagus
Larynx

The stomach

Swallowed food reaches the stomach after being pushed through the esophagus by wavelike muscular contractions called **peristalsis**. Once in the stomach, food mixes with hydrochloric acid and enzymes produced by the stomach lining to begin the digestion of proteins. This lining produces a layer of mucus to protect itself from the acid. The stomach also functions to store partially digested food, **chyme**, for processing later by the small intestine.

Stomach wall

Mucosa
Submucosa
Serosa
Muscularis:
Oblique
Circular
Longitudinal

Simple columnar epithelial cells
Gastric pit
Capillary
Parietal cells *produce hydrochloric acid*
Goblet cells *produce mucus*
Gastric glands

Liver, pancreas & gallbladder

After leaving the stomach, chyme moves into the duodenum, the first part of the small intestine, where it is mixed with bile produced by the **liver** and pancreatic juice produced by the **pancreas**. Bile acts as a mixing agent on the chyme while the pancreatic juice containing numerous digestive enzymes further breaks down fats, proteins and carbohydrates. Excess bile is stored in the **gallbladder**.

Gallbladder
Left hepatic duct
Right hepatic duct
Common hepatic duct
Cystic duct
Common bile duct
Main pancreatic duct
Accessory pancreatic duct
Minor duodenal papilla
Major duodenal papilla
Bile & pancreatic juice
Duodenum
Pancreas

Small intestine

The small intestine consists of three areas: the **duodenum**, **jejunum** and **ileum**. Digestion occurs throughout the entire length of the small intestine, accompanied by the absorption of the resulting molecules by the intestinal wall. **Villi**, projections of the lining of the small intestine, greatly increase the surface area of the absorptive membrane called the **epithelium**. Each cell of the epithelium has microvilli, which further increase this absorptive surface area.

Microvilli
Absorptive epithelial cells

Intestinal lining

Absorbed simple sugars and amino acids pass into capillaries
Absorbed fats pass into lacteals
Epithelium
Villi
Lacteal
Capillary

Layers of large intestine

Longitudinal muscle
Circular muscle
Haustra
Mucosa
Serosa
Teniae coli

Large intestine

The **large intestine** consists of the **cecum**, the **colon** (ascending, transverse, descending and sigmoid) and the **rectum**. As undigested material enters the large intestine, water and electrolytes are absorbed. The remaining waste is stored, formed and expelled.

Free tenia
Ileocecal valve
Semilunar fold
Opening of appendix
Cecum
Appendix

LIVER
GALLBLADDER
DUODENUM
STOMACH
ASCENDING COLON
TRANSVERSE COLON
DESCENDING COLON
JEJUNUM
MESENTERY
ILEUM
RECTUM
SIGMOID COLON

Esophagus
Liver, left lobe
Stomach
Aorta
Portal vein
Celiac trunk
Rugae
Muscularis:
Oblique
Circular
Longitudinal
Superior mesenteric vein & artery

Peristaltic action

Waste material is moved through the digestive system by a series of muscle contractions called **peristalsis**. The contraction of the muscle behind the material moves it into the next section, where the muscle has relaxed.

Rectum & anal canal

Rectum
Inferior rectal valve
Levator ani muscle
Anorectal line
Internal anal sphincter muscle
External anal sphincter muscles:
Deep
Superficial
Subcutaneous
Levator ani muscle

Absorption

Specialized absorptive cells in the epithelium of the small intestine absorb the small molecules produced by digestion. Once absorbed, simple sugars (from carbohydrates) and amino acids (from proteins) enter the capillaries on their way to the portal vein. Absorbed fats enter the lymphatic vessels of the villi, called **lacteals**, before eventually entering the blood.

©Scientific Publishing Ltd., Elk Grove Village, IL USA
#1500

PLATE 24

Understanding GERD
Gastroesophageal Reflux Disease

What is GERD?

GERD (gastroesophageal reflux disease) or **heartburn** is a frequent discomfort. About 1 in 10 adults has heartburn at least once a week; 1 in 3 has the problem at least once a month. Symptoms include a burning sensation in the chest that may start in the upper abdomen and radiate into the neck. Sour or bitter-tasting material is regurgitated into the throat and mouth, especially when lying down or sleeping. Continual chest discomfort after swallowing hard or liquid foods, inflammation of the esophagus, weight loss and vomiting of blood are symptoms of other problems often associated with GERD. Usually a description of symptoms will allow a physician to establish the diagnosis of heartburn.

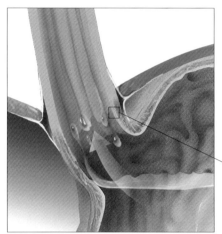

GERD
(Gastroesophageal Reflux Disease)

Under normal circumstances, food passes into the stomach from the esophagus and is prevented from traveling back up the esophagus by the lower esophageal sphincter, which remains tightly closed except when you swallow food. Sometimes, however, the sphincter muscle around the **gastric esophageal junction** becomes weakened and relaxes (opens), allowing acidic stomach contents to move back up the esophagus, producing the symptoms of heartburn.

Gastric esophageal junction

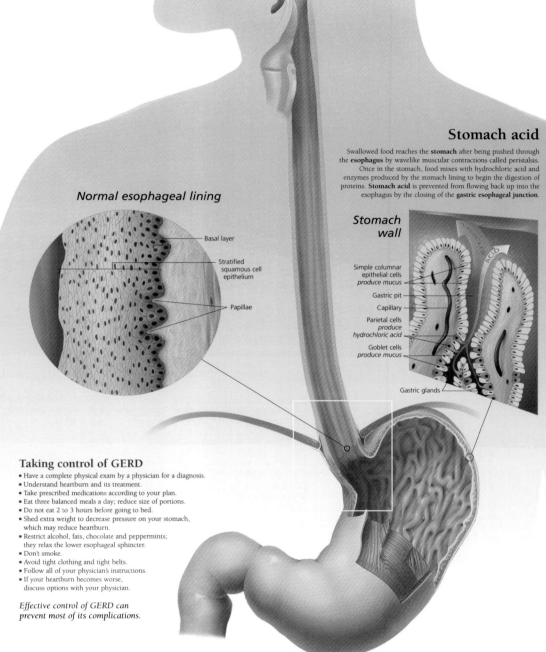

Stomach acid

Swallowed food reaches the **stomach** after being pushed through the **esophagus** by wavelike muscular contractions called peristalsis. Once in the stomach, food mixes with hydrochloric acid and enzymes produced by the stomach lining to begin the digestion of proteins. **Stomach acid** is prevented from flowing back up into the esophagus by the closing of the **gastric esophageal junction**.

Normal esophageal lining

- Basal layer
- Stratified squamous cell epithelium
- Papillae

Stomach wall

- Simple columnar epithelial cells *produce mucus*
- Gastric pit
- Capillary
- Parietal cells *produce hydrochloric acid*
- Goblet cells *produce mucus*
- Gastric glands

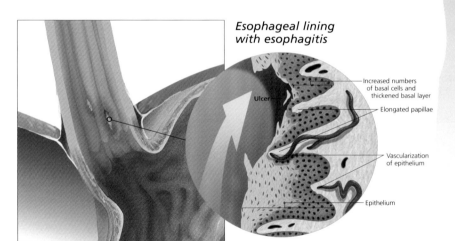

Esophageal lining with esophagitis

- Increased numbers of basal cells and thickened basal layer
- Elongated papillae
- Ulcer
- Vascularization of epithelium
- Epithelium

Esophagitis

When heartburn becomes more frequent, there is a chance of esophagitis, an irritation (inflammation) of the esophageal lining caused by stomach acid. If the esophagitis becomes severe, the result can be bleeding and difficulty in swallowing because of a constriction (stricture) of the esophagus. Some people with severe esophagitis develop Barrett's esophagus.

Taking control of GERD

- Have a complete physical exam by a physician for a diagnosis.
- Understand heartburn and its treatment.
- Take prescribed medications according to your plan.
- Eat three balanced meals a day; reduce size of portions.
- Do not eat 2 to 3 hours before going to bed.
- Shed extra weight to decrease pressure on your stomach, which may reduce heartburn.
- Restrict alcohol, fats, chocolate and peppermints; they relax the lower esophageal sphincter.
- Don't smoke.
- Avoid tight clothing and tight belts.
- Follow all of your physician's instructions.
- If your heartburn becomes worse, discuss options with your physician.

Effective control of GERD can prevent most of its complications.

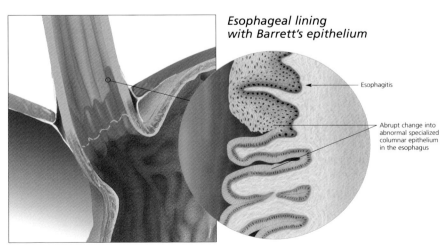

Esophageal lining with Barrett's epithelium

- Esophagitis
- Abrupt change into abnormal specialized columnar epithelium in the esophagus

Barrett's esophagus

In addition to heartburn from a weakened lower esophageal sphincter, many other disorders can result in inflammation of your esophagus. Continual regurgitation of acid from the stomach may damage the normal skin-like lining of the esophagus, which is then replaced by a lining that resembles the lining of the stomach. This new lining usually can resist stomach acid, but inflammation at the upper end of the new lining may narrow (stricture) the interior passageway of the esophagus. Ulcers may occur in the new lining, and can bleed and perforate the esophageal wall. There is a slightly increased risk of cancer occurring in Barrett's esophagus.

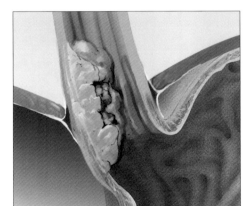

Cancer of the esophagus

Most tumors form in the middle or lower part of the esophagus. The principle symptom of an esophageal tumor is progressive difficulty in swallowing. Beginning with solid foods it will eventually become difficult even to swallow liquids. As the condition worsens, weight loss, the regurgitation of food and foul smelling breath probably will occur. Nearly 90 percent of esophageal tumors are malignant. Any difficulty in swallowing requires immediate attention from a physician for diagnostic tests.

Diet and medication

The ultimate goal of treating heartburn is to produce freedom from symptoms and prevent attacks. Improvements in lifestyle changes and diet alone may be enough to control GERD, especially in mild cases. Often, improvements in diet need to be combined with medication to control GERD. Due to the vast number of mechanisms in the body that affect the digestive system, there are several different types of medications. Your physician will determine which is best for you and may eventually suggest taking more than one.

Acid blockers decrease acid production in your stomach.

Proton Pump Inhibitors are a more powerful inhibitor of stomach acid production and relieves heartburn more effectively.

Surgery is rare and is only a viable solution for those with severe symptoms despite medication and lifestyle changes.

©Scientific Publishing Ltd., Elk Grove Village, IL. USA
#1550

PLATE 25

Understanding IBS
Irritable Bowel Syndrome

What is IBS?

Irritable Bowel Syndrome (IBS) is a functional disorder affecting the large intestine, or colon. In IBS, the colon isn't working properly, leading to chronic and recurrent abdominal discomfort or pain and bowel habit changes. Over the years IBS has been called different names, such as spastic colon and mucous colitis. Irritable Bowel Syndrome has not been shown to lead to other diseases.

Possible causes of IBS

The cause of irritable bowel syndrome is unknown. Since IBS is a functional disorder, there is no infection, inflammation or structural change to be seen. People with IBS seem to have increased colon sensitivity. Their colon muscles spasm, or suddenly contract, after only mild stimulation. This increased gut sensitivity may result from a problem in the interaction of the brain, the intestines and the autonomic nervous system. The most likely triggers of IBS symptoms are diet, emotional stress and hormonal changes. IBS may, in some cases, be the result of another disease.

Colon spasm (external)

Normally muscles contract and relax in a coordinated rhythm to move waste material through the colon. In IBS, this rhythm is interrupted by a sudden, involuntary contraction called a spasm.

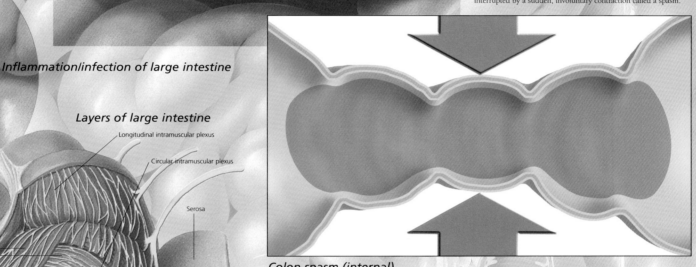

Inflammation/infection of large intestine

Layers of large intestine

Longitudinal intramuscular plexus
Circular intramuscular plexus
Serosa
Haustra
Longitudinal muscle
Circular muscle
Mucosa
Teniae coli

Colon spasm (internal)

Symptoms

Normal bowel function varies from one person to the next, and most people have a bowel disturbance from time to time. People with IBS may experience chronic and recurrent abdominal discomfort or pain and bowel disturbances, such as diarrhea, constipation or alternating diarrhea and constipation. Other symptoms may include:

- Change in frequency of bowel movements
- Abdominal pain relieved by defecation
- Bloating
- Excessive amount of gas
- Passage of mucous with a bowel movement
- Feeling that the bowel is not completely empty
- Nausea

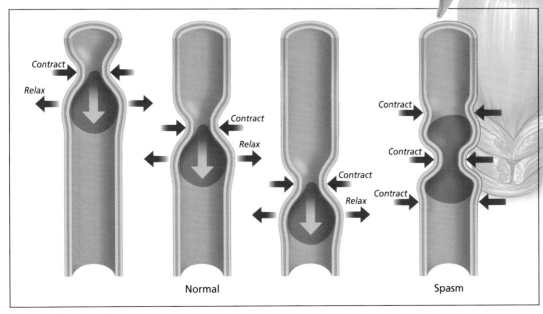

Contract
Relax
Contract
Relax
Contract
Relax
Contract
Contract
Contract
Contract

Normal
Spasm

Peristaltic action

Waste material is moved through the colon by a series of muscle contractions called peristalsis. The contraction of the muscle behind the material moves it into the next section of colon, where the muscle has relaxed. In IBS, spasms interrupt the process, producing discomfort or pain and bowel disturbances.

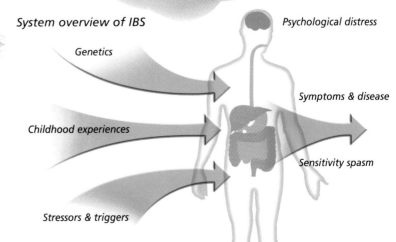

System overview of IBS

Psychological distress

Genetics

Symptoms & disease

Childhood experiences

Sensitivity spasm

Stressors & triggers

Symptom management

- Evaluate your diet to see if there are foods that appear to trigger IBS symptoms
- Increase fiber intake as this softens stool and improves movement through the intestinal tract
- Examine external factors such as home, work or financial burdens to identify stresses
- Consult a doctor about medicines such as antispasmodics or analgesics

©Scientific Publishing Ltd., Elk Grove Village, IL. USA
#1551

PLATE 26

Diseases of the Digestive System

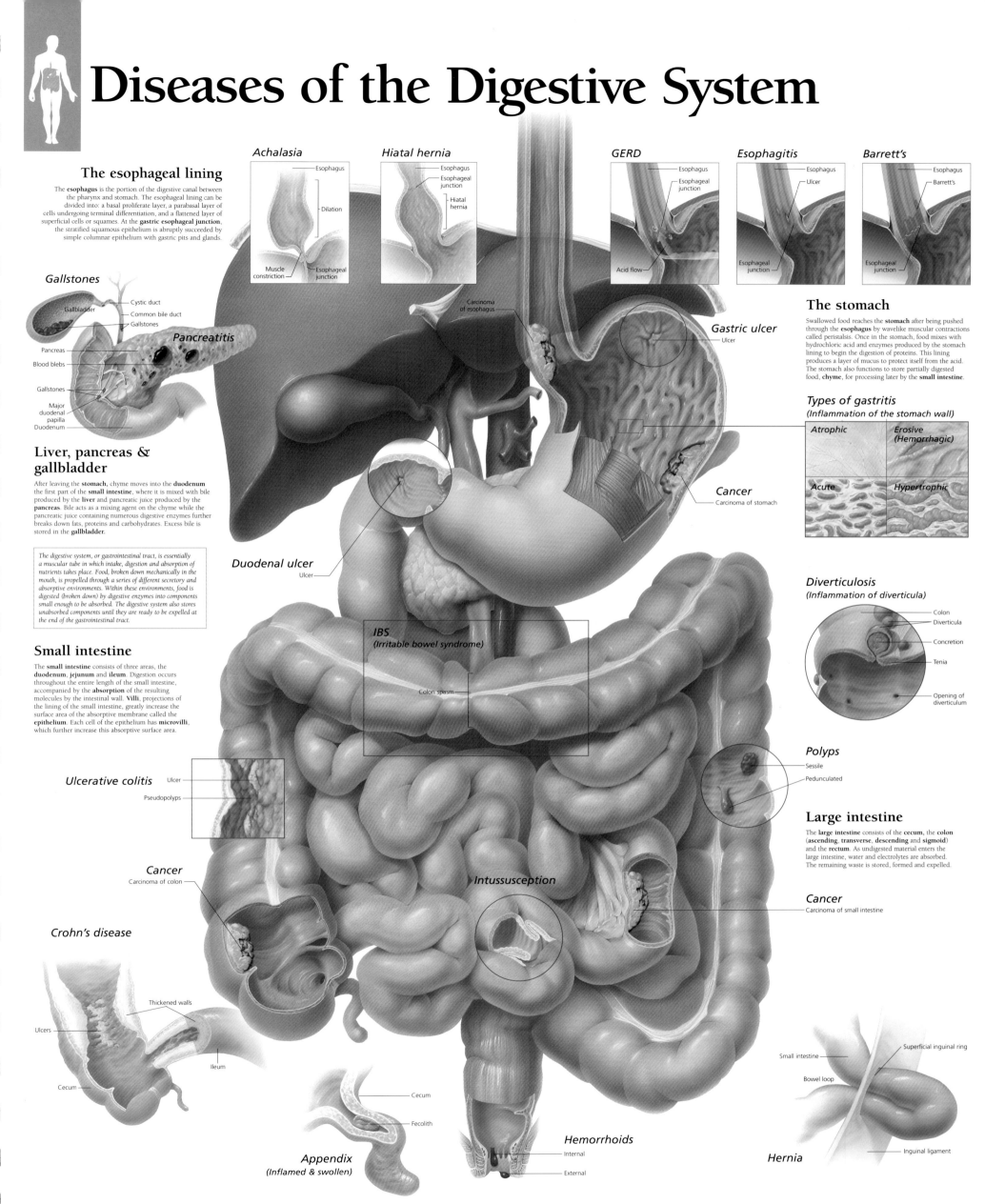

The esophageal lining

The **esophagus** is the portion of the digestive canal between the pharynx and stomach. The esophageal lining can be divided into: a basal proliferate layer, a parabasal layer of cells undergoing terminal differentiation, and a flattened layer of superficial cells or squames. At the **gastric esophageal junction**, the stratified squamous epithelium is abruptly succeeded by simple columnar epithelium with gastric pits and glands.

Achalasia
- Esophagus
- Dilation
- Muscle constriction
- Esophageal junction

Hiatal hernia
- Esophagus
- Esophageal junction
- Hiatal hernia

GERD
- Esophagus
- Esophageal junction
- Acid flow

Esophagitis
- Esophagus
- Ulcer
- Esophageal junction

Barrett's
- Esophagus
- Barrett's
- Esophageal junction

Gallstones
- Gallbladder
- Cystic duct
- Common bile duct
- Gallstones

Pancreatitis
- Pancreas
- Blood blebs
- Gallstones
- Major duodenal papilla
- Duodenum

Liver, pancreas & gallbladder

After leaving the **stomach**, chyme moves into the **duodenum** the first part of the **small intestine**, where it is mixed with bile produced by the **liver** and pancreatic juice produced by the **pancreas**. Bile acts as a mixing agent on the chyme while the pancreatic juice containing numerous digestive enzymes further breaks down fats, proteins and carbohydrates. Excess bile is stored in the **gallbladder**.

The digestive system, or gastrointestinal tract, is essentially a muscular tube in which intake, digestion and absorption of nutrients takes place. Food, broken down mechanically in the mouth, is propelled through a series of different secretory and absorptive environments. Within these environments, food is digested (broken down) by digestive enzymes into components small enough to be absorbed. The digestive system also stores unabsorbed components until they are ready to be expelled at the end of the gastrointestinal tract.

Small intestine

The **small intestine** consists of three areas, the **duodenum**, **jejunum** and **ileum**. Digestion occurs throughout the entire length of the small intestine, accompanied by the **absorption** of the resulting molecules by the intestinal wall. **Villi**, projections of the lining of the small intestine, greatly increase the surface area of the absorptive membrane called the **epithelium**. Each cell of the epithelium has **microvilli**, which further increase this absorptive surface area.

Carcinoma of esophagus

Gastric ulcer
- Ulcer

Cancer
- Carcinoma of stomach

Duodenal ulcer
- Ulcer

IBS
(Irritable bowel syndrome)
- Colon spasm

Ulcerative colitis
- Ulcer
- Pseudopolyps

Cancer
- Carcinoma of colon

Intussusception

Crohn's disease
- Thickened walls
- Ulcers
- Ileum
- Cecum

The stomach

Swallowed food reaches the **stomach** after being pushed through the **esophagus** by wavelike muscular contractions called peristalsis. Once in the stomach, food mixes with hydrochloric acid and enzymes produced by the stomach lining to begin the digestion of proteins. This lining produces a layer of mucus to protect itself from the acid. The stomach also functions to store partially digested food, **chyme**, for processing later by the **small intestine**.

Types of gastritis
(Inflammation of the stomach wall)

Atrophic	Erosive (Hemorrhagic)
Acute	Hypertrophic

Diverticulosis
(Inflammation of diverticula)
- Colon
- Diverticula
- Concretion
- Tenia
- Opening of diverticulum

Polyps
- Sessile
- Pedunculated

Large intestine

The **large intestine** consists of the **cecum**, the **colon** (**ascending**, **transverse**, **descending** and **sigmoid**) and the **rectum**. As undigested material enters the large intestine, water and electrolytes are absorbed. The remaining waste is stored, formed and expelled.

Cancer
- Carcinoma of small intestine

Appendix
(Inflamed & swollen)
- Cecum
- Fecolith

Hemorrhoids
- Internal
- External

Hernia
- Small intestine
- Bowel loop
- Superficial inguinal ring
- Inguinal ligament

©Scientific Publishing Ltd., Elk Grove Village, IL, USA
#1552

PLATE 27

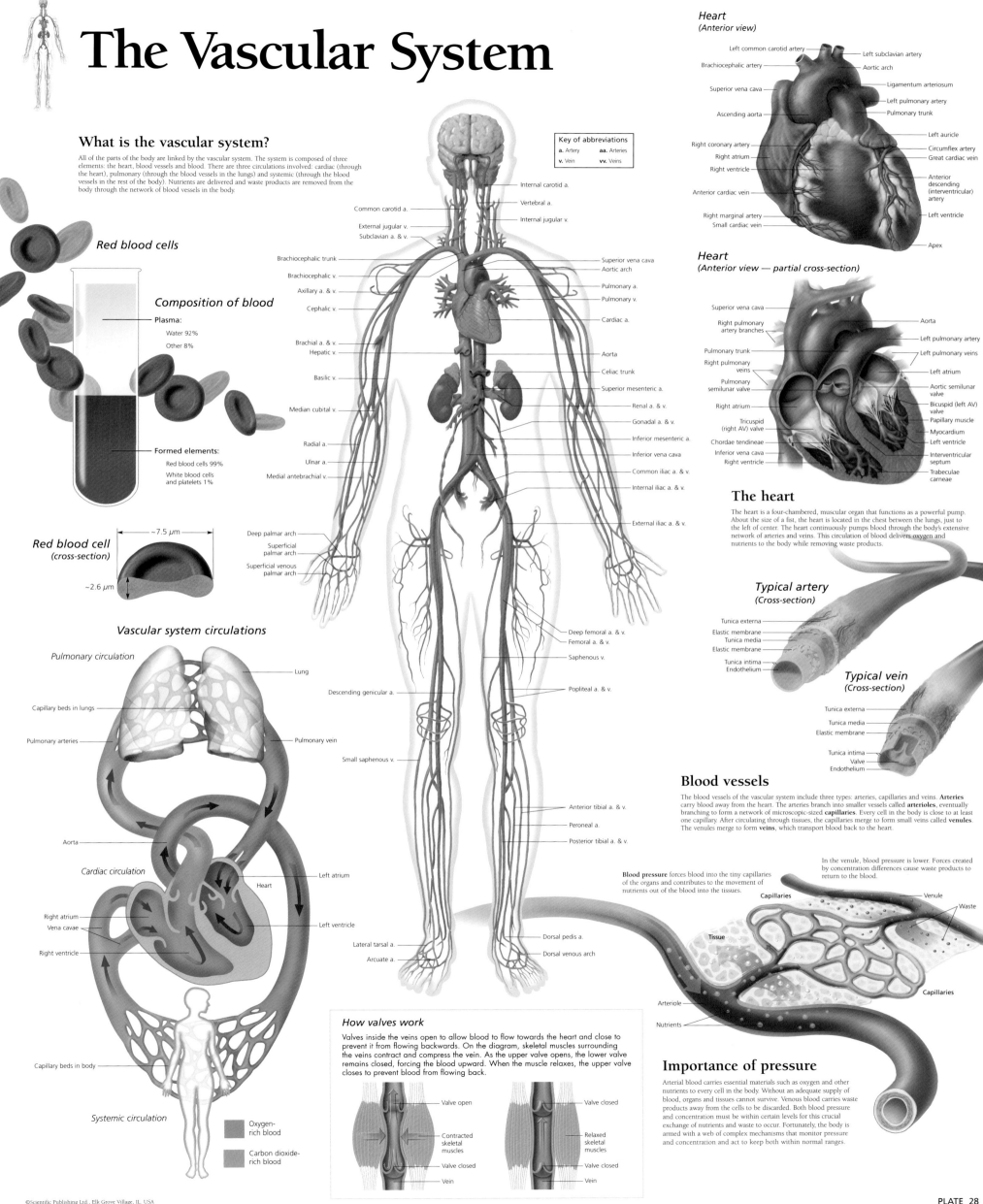

The Vascular System

Heart
(Anterior view)

Left common carotid artery
Brachiocephalic artery
Superior vena cava
Ascending aorta
Right coronary artery
Right atrium
Right ventricle
Anterior cardiac vein
Right marginal artery
Small cardiac vein
Left subclavian artery
Aortic arch
Ligamentum arteriosum
Left pulmonary artery
Pulmonary trunk
Left auricle
Circumflex artery
Great cardiac vein
Anterior descending (interventricular) artery
Left ventricle
Apex

What is the vascular system?

All of the parts of the body are linked by the vascular system. The system is composed of three elements: the heart, blood vessels and blood. There are three circulations involved: cardiac (through the heart), pulmonary (through the blood vessels in the lungs) and systemic (through the blood vessels in the rest of the body). Nutrients are delivered and waste products are removed from the body through the network of blood vessels in the body.

Key of abbreviations
a. Artery   **aa.** Arteries
v. Vein   **vv.** Veins

Red blood cells

Composition of blood

Plasma:
Water 92%
Other 8%

Formed elements:
Red blood cells 99%
White blood cells and platelets 1%

Red blood cell
(cross-section)

~7.5 μm
~2.6 μm

Internal carotid a.
Common carotid a.
External jugular v.
Subclavian a. & v.
Vertebral a.
Internal jugular v.
Brachiocephalic trunk
Brachiocephalic v.
Axillary a. & v.
Cephalic v.
Brachial a. & v.
Hepatic v.
Basilic v.
Median cubital v.
Radial a.
Ulnar a.
Medial antebrachial v.
Superior vena cava
Aortic arch
Pulmonary a.
Pulmonary v.
Cardiac a.
Aorta
Celiac trunk
Superior mesenteric a.
Renal a. & v.
Gonadal a. & v.
Inferior mesenteric a.
Inferior vena cava
Common iliac a. & v.
Internal iliac a. & v.
External iliac a. & v.

Deep palmar arch
Superficial palmar arch
Superficial venous palmar arch

Deep femoral a. & v.
Femoral a. & v.
Saphenous v.
Popliteal a. & v.
Descending genicular a.
Small saphenous v.
Anterior tibial a. & v.
Peroneal a.
Posterior tibial a. & v.
Lateral tarsal a.
Arcuate a.
Dorsal pedis a.
Dorsal venous arch

Heart
(Anterior view — partial cross-section)

Superior vena cava
Right pulmonary artery branches
Pulmonary trunk
Right pulmonary veins
Pulmonary semilunar valve
Right atrium
Tricuspid (right AV) valve
Chordae tendineae
Inferior vena cava
Right ventricle
Aorta
Left pulmonary artery
Left pulmonary veins
Left atrium
Aortic semilunar valve
Bicuspid (left AV) valve
Papillary muscle
Myocardium
Left ventricle
Interventricular septum
Trabeculae carneae

The heart

The heart is a four-chambered, muscular organ that functions as a powerful pump. About the size of a fist, the heart is located in the chest between the lungs, just to the left of center. The heart continuously pumps blood through the body's extensive network of arteries and veins. This circulation of blood delivers oxygen and nutrients to the body while removing waste products.

Typical artery
(Cross-section)

Tunica externa
Elastic membrane
Tunica media
Elastic membrane
Tunica intima
Endothelium

Typical vein
(Cross-section)

Tunica externa
Tunica media
Elastic membrane
Tunica intima
Valve
Endothelium

Blood vessels

The blood vessels of the vascular system include three types: arteries, capillaries and veins. **Arteries** carry blood away from the heart. The arteries branch into smaller vessels called **arterioles**, eventually branching to form a network of microscopic-sized **capillaries**. Every cell in the body is close to at least one capillary. After circulating through tissues, the capillaries merge to form small veins called **venules**. The venules merge to form **veins**, which transport blood back to the heart.

Vascular system circulations

Pulmonary circulation

Lung
Capillary beds in lungs
Pulmonary arteries
Pulmonary vein

Aorta

Cardiac circulation

Heart
Right atrium
Vena cavae
Right ventricle
Left atrium
Left ventricle

Capillary beds in body

Systemic circulation

Oxygen-rich blood
Carbon dioxide-rich blood

Blood pressure forces blood into the tiny capillaries of the organs and contributes to the movement of nutrients out of the blood into the tissues.

In the venule, blood pressure is lower. Forces created by concentration differences cause waste products to return to the blood.

Capillaries
Venule
Waste
Tissue
Capillaries
Arteriole
Nutrients

How valves work

Valves inside the veins open to allow blood to flow towards the heart and close to prevent it from flowing backwards. On the diagram, skeletal muscles surrounding the veins contract and compress the vein. As the upper valve opens, the lower valve remains closed, forcing the blood upward. When the muscle relaxes, the upper valve closes to prevent blood from flowing back.

Valve open
Contracted skeletal muscles
Valve closed
Vein

Valve closed
Relaxed skeletal muscles
Valve closed
Vein

Importance of pressure

Arterial blood carries essential materials such as oxygen and other nutrients to every cell in the body. Without an adequate supply of blood, organs and tissues cannot survive. Venous blood carries waste products away from the cells to be discarded. Both blood pressure and concentration must be within certain levels for this crucial exchange of nutrients and waste to occur. Fortunately, the body is armed with a web of complex mechanisms that monitor pressure and concentration and act to keep both within normal ranges.

©Scientific Publishing Ltd., Elk Grove Village, IL. USA
#1600

PLATE 28

Understanding Diabetes

Glucose metabolism

After a meal, carbohydrates are converted to glucose by the digestive system. The glucose then enters the bloodstream. The pancreas responds to the rise in the blood glucose level by producing insulin and secreting it into the bloodstream. Insulin has several effects—it suppresses glucose production in the liver, it signals the liver to increase glucose uptake, and it allows glucose to enter cells from the bloodstream. As a result of these actions, blood glucose levels fall, insulin production ceases, and homeostasis is restored.

When the blood glucose level drops too far (such as after skipping a meal), the pancreas produces glucagon and secretes it into the bloodstream. Glucagon signals the liver to change glycogen into glucose and to release the glucose into the bloodstream. As blood glucose levels rise, glucagon production ceases, and homeostasis is restored.

Glucose

Insulin stimulates cells to take up glucose

Glucose   Glycogen   *Liver*

Glucagon signals the liver to convert glycogen to glucose and release the glucose into the blood

Pancreas

What is diabetes?

Diabetes is a disease characterized by a chronic imbalance in the blood glucose levels. There are two primary types of diabetes. Diabetes can strike at any age, although historically, young people normally fall victim because insulin production completely ceases. Those afflicted at later ages have some insulin production, but not enough to maintain a healthy blood glucose level. There are also two lesser-known types of diabetes. One affects women during pregnancy and is termed "gestational diabetes." It most often disappears or subsides substantially after the pregnancy ends. The fourth type of diabetes is the result of pancreatic disease, hormonal irregularities, or harmful drug interaction.

Insulin

Blood glucose level

Glucagon
Epinephrine cortisol
Human growth hormone

Cells use glucose as fuel

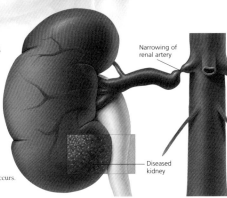

Glucose is converted to triglycerides and stored in fat (adipose tissues)

The liver takes up glucose, converts it into glycogen and stores it

Function of insulin

Insulin, which is produced by the pancreas, facilitates the absorption of glucose into muscles for fuel. Glucose provides power to the body, but is kept under control by insulin. When the body acts normally, insulin bonds to the surface of cells and, as glucose travels throughout the body, glucose is able to penetrate the cell and be used effectively. Insulin determines how much glucose is produced by the liver and between meals. It does this by countering another hormone, glucagon, also produced by the pancreas. Glucagon sends a message to the liver to convert glycogen to glucose.

Action of insulin

Type I

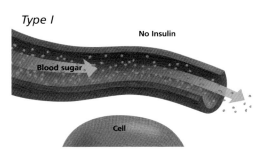

No Insulin

Blood sugar

Cell

Type I diabetes is less prevalent and occurs when there is a complete failure of the pancreas to produce insulin. It is considered an autoimmune disease. Type I diabetes occurs when antibodies produced by the body attack beta cells. Beta cells, which secrete insulin, are damaged over time and the cells eventually cease to produce the needed insulin. The absence of insulin allows ketones, normally a beneficial substance produced by the liver, to build up to abnormal levels, causing an acidosis (diabetic) coma.

Type II

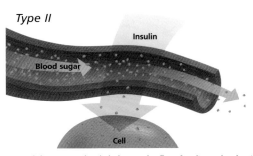

Insulin

Blood sugar

Cell

Type II diabetes occurs when the body resists the effects of insulin, or when there is a partial failure of the pancreas to produce insulin. Overweight people are often afflicted with this type of diabetes because their tissue becomes unable to respond to insulin and the pancreas is forced to produce more, but cannot meet the demand. The lack of insulin prevents glucose from entering the cells, so the body is not fully energized. This type of diabetes is normally diagnosed later in life as the continual strain on the pancreas leads to a decline of insulin production.

Normal

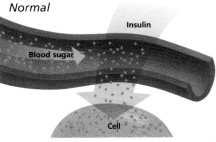

Insulin

Blood sugar

Cell

The interaction of glucose and insulin maintains a **normal level** of blood sugar and enables the body to perform daily activities.

Symptoms of diabetes

- Increased frequency of urination
- Unusually high desire for fluids
- Weight loss
- Blurred vision
- Weakness and fatigue
- Skin infections
- Complications that may include vascular disease and nerve damage

Health complications

Kidney damage

The kidney's ability to filter waste from the bloodstream is hindered by diabetes. High glucose levels, plus the common side effect of high blood pressure, damage the group of capillaries within the kidney called the glomeruli. A normal kidney filters waste and discharges it through the urine, allowing the necessary proteins to flow back through the body. Kidney damage, termed "nephropathy," is irreversible, leaving dialysis or a kidney transplant as the only two means to replace vital kidney processes once complete kidney failure occurs.

Narrowing of renal artery

Diseased kidney

Vision problems

Vision problems are a common side effect of diabetes, specifically a condition known as retinopathy. In this eye complication, the small blood vessels that supply the retina with blood weaken and leak, damaging the retina and hindering its ability to transmit images to the brain. A more severe condition arises when fragile blood vessels grow into the vitreous humor (a jelly-like substance at the back of the eyeball) and rupture. Diabetics also have an increased risk for glaucoma and cataracts.

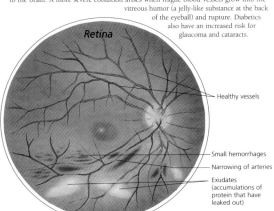

Retina

Healthy vessels

Small hemorrhages

Narrowing of arteries

Exudates (accumulations of protein that have leaked out)

Heart disease

Blocked coronary arteries

Enlarged heart
Normal heart
Damaged heart tissue

Atherosclerosis, or narrowing of the arteries, may cause circulation to deteriorate in the coronary arteries. This leads to an increased possibility of angina, pain caused by a reduced blood flow to the heart, or heart attack. Strokes are also more prevalent in diabetics, as arteries to the brain are also affected by reduced blood flow.

Nerve damage

Diabetics' higher glucose levels make them more susceptible to nerve damage. The most common form is peripheral neuropathy, which causes limbs to tingle as a result of the reduced function of sensory nerves. Peripheral neuropathy develops slowly and creates numbness and pain in the hands and legs. Other forms of nerve damage may also affect the digestive tract, bladder and other internal organs.

Treatment and healthy lifestyle changes

Proper diabetes control depends on glucose levels. When glucose levels are too low, diabetics occasionally experience hypoglycemia, or low blood sugar. Hypoglycemia typically occurs when the insulin dosage exceeds the amount needed, such as between meals. The opposite condition is hyperglycemia, resulting when the body is not able to burn off sugars through physical activity. Both conditions occur often in diabetics because glucose levels controlled through medication cannot be as precisely regulated as through the body's natural mechanisms.

The successful treatment for diabetes, a non-curable disease, requires that patients maintain a healthy lifestyle through diet, medication and exercise. Here are a few suggestions:

1) Monitor blood glucose levels. Seek the advice of a doctor or nurse as to the type (blood or urine) and the frequency of tests required.

2) Eat regularly and do not skip meals. This advice is very important to maintain proper blood glucose levels. The amount of food eaten must be balanced with the amount of energy expended. If the patient is taking insulin, regular consultation with a dietician or nurse is recommended to help coordinate the timing of injections with meals.

3) Choose the right foods. Balance starchy, high-fiber foods with vegetables, fruits, and proteins. Avoid sweets and high-fat foods. Limit salt intake.

4) Limit or avoid alcohol. For patients on medication, whether orally or by injection, alcohol can lead to a hypoglycemic attack.

5) Exercise regularly.

6) Maintain ideal body weight. Lose weight if necessary by eating a well-balanced diet. Avoid fad diets.

7) Take medication as prescribed.

©Scientific Publishing Ltd., Elk Grove Village, IL. USA
#1650

PLATE 29

Understanding Cholesterol

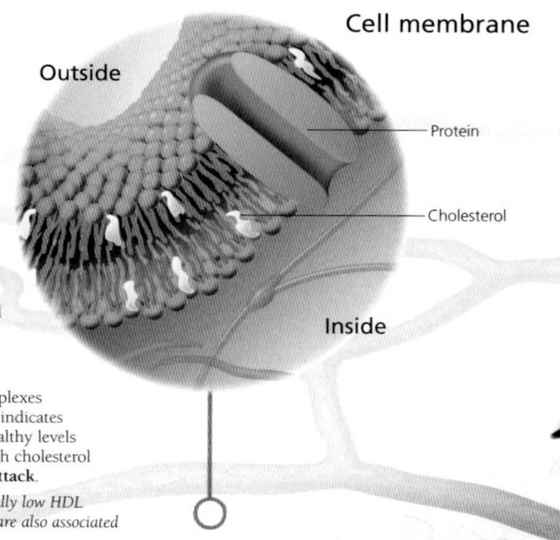

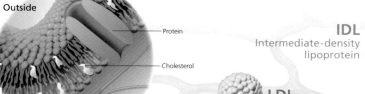

How cholesterol travels

High-density lipoproteins carry excess cholesterol away from cells for reprocessing in the liver or elimination through the digestive system. This mechanism protects against high cholesterol as well as atherosclerosis or "hardening of the arteries." When HDL levels are too low, less cholesterol is removed from the tissues and cholesterol levels increase.

Low-density lipoproteins carry cholesterol produced in the liver to tissues throughout the body. High concentrations of LDL are a leading factor in the development of atherosclerosis.

Very low-density lipoproteins, consisting primarily of cholesterol with little protein, also carry cholesterol to the body's tissues and can deposit cholesterol on blood vessel walls.

What is cholesterol?

Cholesterol is a natural, fat-like substance that is indispensable to the human body. It is a key component of cells, helping to maintain the stability and fluidity of **cell membranes**. It is utilized abundantly in the liver to produce **cholic acid**, which forms the bile salts necessary for **fat digestion**. Cholesterol is also necessary for the formation of hormones such as **estrogen** and **testosterone**. Significant amounts of cholesterol are used in the skin to synthesize **Vitamin D** and to help control **water evaporation** through the pores.

The body's cholesterol supply comes from two sources. It is primarily manufactured in the **liver**, along with several other organs in the body. It is also ingested through food, particularly **eggs**, **red meat**, and **dairy products** high in cholesterol content.

What does "high cholesterol" mean?

Blood cholesterol tests measure the amount of cholesterol bound to **lipoproteins**, fat-protein complexes that carry fats through the bloodstream. A diagnosis of high cholesterol or **hypercholesterolemia** indicates total cholesterol levels of 240 mg/dL or above (*see chart below*). More specifically, it indicates unhealthy levels of **low-density lipoproteins (LDL)**, **high-density lipoproteins (HDL)**, and/or **triglycerides**. High cholesterol caused by any of these factors increases the risk of **coronary artery disease** and potential **heart attack**.

High cholesterol is often thought of as an excess amount of LDL ("bad") cholesterol. However, an abnormally low HDL ("good") cholesterol level is an equally important risk factor for heart disease. Elevated triglyceride levels are also associated with increased risk, especially in combination with obesity and other factors.

Cholesterol guidelines (*see chart below*) are used to identify appropriate cholesterol levels for different individuals. Another valuable tool for predicting coronary artery disease is a **risk ratio**. This calculation measures the ratio of one form of cholesterol to another by dividing total cholesterol by either the HDL or LDL level.

- Optimal risk ratio – 3:5
- Average risk ratio – 4:5
- Increased risk ratio – 5:1

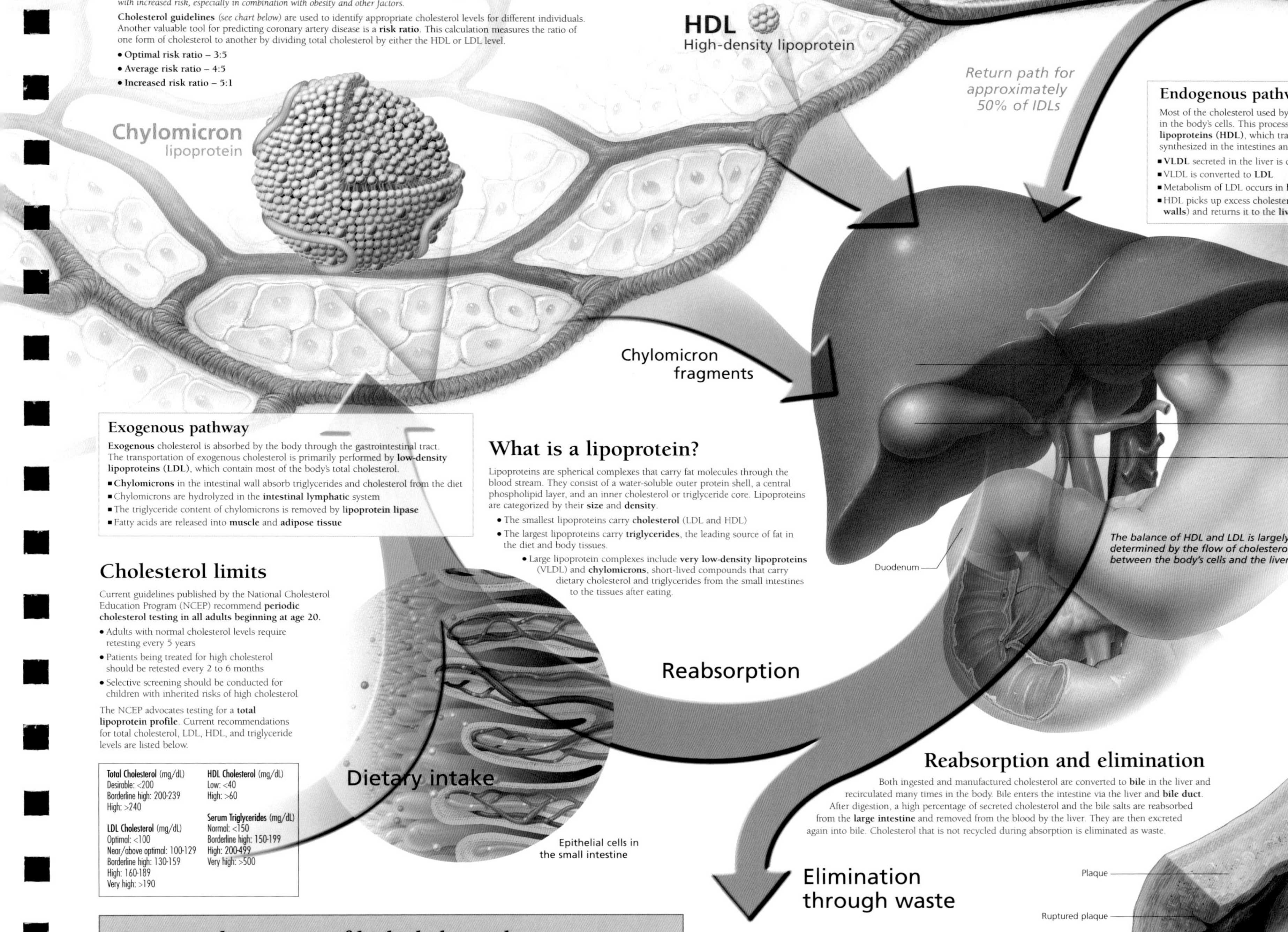

Cholesterol limits

Current guidelines published by the National Cholesterol Education Program (NCEP) recommend **periodic cholesterol testing in all adults beginning at age 20.**

- Adults with normal cholesterol levels require retesting every 5 years
- Patients being treated for high cholesterol should be retested every 2 to 6 months
- Selective screening should be conducted for children with inherited risks of high cholesterol

The NCEP advocates testing for a **total lipoprotein profile**. Current recommendations for total cholesterol, LDL, HDL, and triglyceride levels are listed below.

Total Cholesterol (mg/dL)	HDL Cholesterol (mg/dL)
Desirable: <200	Low: <40
Borderline high: 200-239	High: >60
High: >240	
LDL Cholesterol (mg/dL)	**Serum Triglycerides (mg/dL)**
Optimal: <100	Normal: <150
Near/above optimal: 100-129	Borderline high: 150-199
Borderline high: 130-159	High: 200-499
High: 160-189	Very high: >500
Very high: >190	

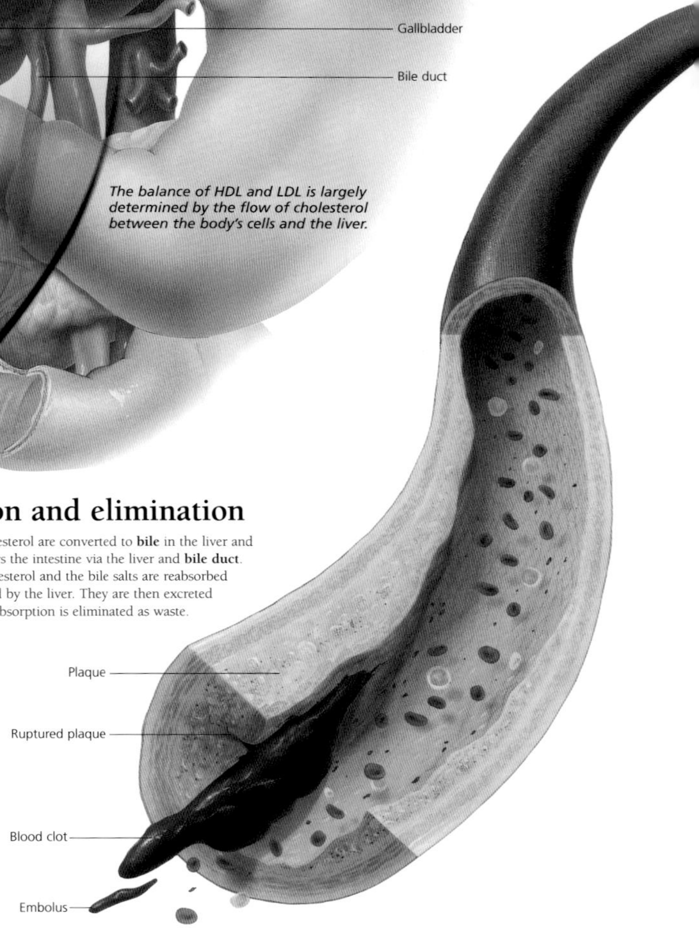

Causes and treatment of high cholesterol

Causes
High cholesterol is caused by a number of factors that can be controlled to prevent or reduce the risk of heart disease.

High saturated fat and cholesterol intake increase total blood cholesterol by increasing production of cholesterol in the liver and slowing cholesterol elimination from the body.

Being overweight elevates cholesterol levels and increases the risk of heart disease.

A sedentary lifestyle increases LDL levels and decreases protective levels of HDL.

Smoking also increases heart disease risks by reducing protective HDL.

Age, gender and heredity also play a role in elevated cholesterol. These factors can make cholesterol levels more difficult to control.

Treatment
Lowering cholesterol can require a combination of diet changes, increased exercise, and medication.

Dietary modification is the most important step in aggressively treating high cholesterol. Recommendations include:
- a maximum 7% daily dietary intake of saturated fats
- replacing saturated fats with unsaturated fats such as olive or canola oil
- a maximum 200 mg/day of cholesterol intake
- increased consumption of fiber-rich foods including whole grains and fruits and vegetables

Regular physical activity is also essential to:
- lower LDL levels
- increase HDL levels
- promote weight loss
- reduce the risk of atherosclerosis

Medication therapy is used to reduce cholesterol when diet and exercise are not sufficient. Lipid-lowering drugs can:
- improve the balance of HDL and LDL
- reduce serum triglyceride levels

What is a lipoprotein?

Lipoproteins are spherical complexes that carry fat molecules through the blood stream. They consist of a water-soluble outer protein shell, a central phospholipid layer, and an inner cholesterol or triglyceride core. Lipoproteins are categorized by their **size** and **density**.

- The smallest lipoproteins carry **cholesterol** (LDL and HDL)
- The largest lipoproteins carry **triglycerides**, the leading source of fat in the diet and body tissues
- Large lipoprotein complexes include **very low-density lipoproteins (VLDL)** and **chylomicrons**, short-lived compounds that carry dietary cholesterol and triglycerides from the small intestines to the tissues after eating.

Exogenous pathway

Exogenous cholesterol is absorbed by the body through the gastrointestinal tract. The transportation of exogenous cholesterol is primarily performed by **low-density lipoproteins (LDL)**, which contain most of the body's total cholesterol.

- **Chylomicrons** in the intestinal wall absorb triglycerides and cholesterol from the diet
- Chylomicrons are hydrolyzed in the **intestinal lymphatic** system
- The triglyceride content of chylomicrons is removed by **lipoprotein lipase**
- Fatty acids are released into **muscle** and **adipose tissue**

Endogenous pathway

Most of the cholesterol used by the body is **endogenous**, formed in the body's cells. This process primarily involves **high-density lipoproteins (HDL)**, which transport endogenous cholesterol synthesized in the intestines and other organs.

- VLDL secreted in the liver is carried to muscle and adipose tissue
- VLDL is converted to **LDL**
- Metabolism of LDL occurs in liver and other cells
- HDL picks up excess cholesterol in the cells (including **artery walls**) and returns it to the **liver** for disposal

Reabsorption and elimination

Both ingested and manufactured cholesterol are converted to **bile** in the liver and recirculated many times in the body. Bile enters the intestine via the liver and **bile duct**. After digestion, a high percentage of secreted cholesterol and the bile salts are reabsorbed from the **large intestine** and removed from the blood by the liver. They are then excreted again into bile. Cholesterol that is not recycled during absorption is eliminated as waste.

What is atherosclerosis?

Atherosclerosis is the gradual buildup of plaque in the arteries, caused primarily by **low-density lipoprotein (LDL)** deposited inside the vessel walls. Localized plaques or atheromas made up of fats and cholesterol thicken and eventually protrude into the artery. While small **atheromas** often remain soft, older atheromas may become larger and develop fibrous calcium deposits on the surface. During progression of the disease, arteries become increasingly calcified and inelastic, reducing or blocking blood flow and preventing oxygen-rich blood from reaching the heart. Atheromas present an additional danger if they become sites for blood clots, which may rupture and result in a heart attack.

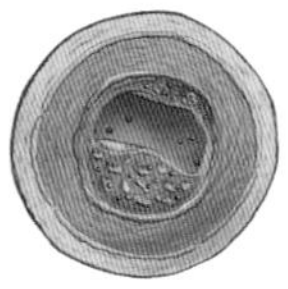

Progression of plaque development

- Oxidized low-density lipoproteins initiate endothelial cell injury
- Fatty streaks consisting of lipid-filled macrophages and lymphocytes appear
- Layers of macrophages and smooth muscle are present
- Lesions or fibrous plaques develop over accumulated lipids and debris, protruding into the artery

Labels and captions

Cell membrane

Outside

Inside

Protein

Cholesterol

IDL — Intermediate-density lipoprotein

LDL — Low-density lipoprotein

VLDL — Very low-density lipoprotein

HDL

HDL — High-density lipoprotein

Chylomicron lipoprotein

Return path for approximately 50% of IDLs

Chylomicron fragments

Liver

Gallbladder

Bile duct

Duodenum

The balance of HDL and LDL is largely determined by the flow of cholesterol between the body's cells and the liver.

Reabsorption

Dietary intake

Epithelial cells in the small intestine

Elimination through waste

Plaque

Ruptured plaque

Blood clot

Embolus

Normal artery

Artery with plaque

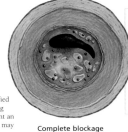

Complete blockage

©Scientific Publishing Ltd., Elk Grove Village, IL, USA
#1651

PLATE 30

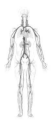

Understanding DVT
Deep Vein Thrombosis

What is the venous system?

The venous system returns deoxygenated blood and waste CO_2 to the heart and lungs. The body's veins originate as capillaries in the organs and tissues and branch into venules, eventually joining to become veins. Superficial veins lie closer to the skin surface. Deep veins, where DVTs usually occur, are surrounded by muscles, which help to push blood towards the heart as they contract. These larger veins also contain many one-way valves to prevent blood from pooling and flowing backwards. Veins have thinner walls and fewer elastic fibers than arteries, which transport blood from the heart under significantly higher pressures.

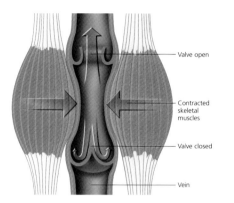

- Valve open
- Contracted skeletal muscles
- Valve closed
- Vein

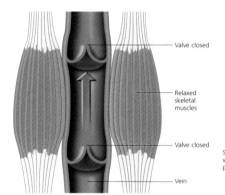

- Valve closed
- Relaxed skeletal muscles
- Valve closed
- Vein

How valves work

Valves inside the veins open to allow blood to flow towards the heart and close to prevent it from flowing backwards. In the diagram above, skeletal muscles surrounding the veins contract and compress the vein. As the upper valve opens, the lower valve remains closed, forcing the blood upward. When the muscle relaxes, the upper valve closes to prevent blood from flowing back.

What is DVT?

DVT or Deep Vein Thrombosis is a blood clot that forms in a deep vein of the body, most commonly in the legs. The danger increases if the clot loosens and is carried by the blood to the lungs, causing a blockage in the lung arteries. This can result in death or serious long-term health complications.

DVT is often due to inactivity of the body's lower extremities. Most commonly, it develops in those who are bedridden or have been unable to move for long periods of time, such as during an overseas flight. Other risks include specific health conditions, surgeries and medications.

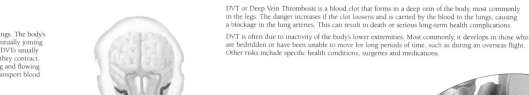
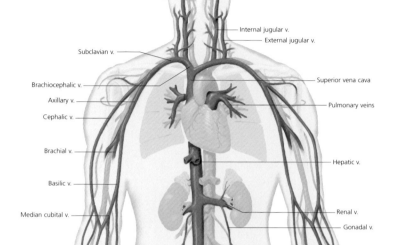

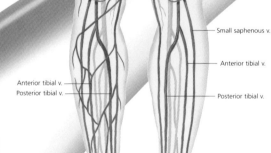

- Internal jugular v.
- External jugular v.
- Subclavian v.
- Superior vena cava
- Brachiocephalic v.
- Pulmonary veins
- Axillary v.
- Cephalic v.
- Brachial v.
- Hepatic v.
- Basilic v.
- Median cubital v.
- Renal v.
- Gonadal v.
- Inferior vena cava
- Median antebrachial v.
- Common iliac v.
- Internal iliac v.
- External iliac v.
- Superficial venous palmar arch
- Deep femoral v.
- Femoral v.
- Common iliac v.
- Deep femoral v.
- Great saphenous v.
- Femoral v.
- Great saphenous v.
- Popliteal v.
- Popliteal v.
- Small saphenous v.
- Anterior tibial v.
- Anterior tibial v.
- Posterior tibial v.
- Posterior tibial v.
- Dorsal venous arch

Superficial and deep veins   *Deep veins*

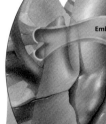

- Acute infarct
- Embolism

What is a pulmonary embolism?

Pulmonary embolisms occur when foreign material, frequently a blood clot, lodges in an artery in the lungs. Blood flow is slowed or stopped, causing sharp, sudden chest pain that becomes more severe when taking a deep breath. The illustration (above) indicates an embolism occurring in the left pulmonary artery with acute infarct (a sudden insufficiency in the blood supply).

Chronic Venous Insufficiency

CVI or Chronic Venous Insufficiency is a condition affecting the veins of the lower legs. It typically occurs as a result of damage to the valves in the veins, often specifically by DVTs. These valves normally prevent blood from flowing backwards, but become weak and incompetent, allowing blood to pool, increasing venous pressure and resulting in swelling, skin ulcers, yellow discoloration and tissue loss. CVI may also result when the muscles surrounding the vein do not contract properly and blood cannot be efficiently directed upwards to the heart.

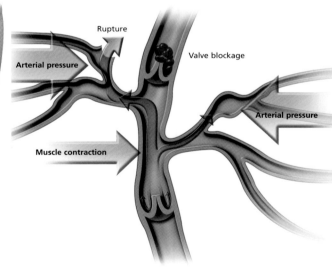

- Rupture
- Valve blockage
- Arterial pressure
- Arterial pressure
- Muscle contraction

This illustration demonstrates changes in the leg when CVI is present. A condition known as stasis develops because the veins and leg muscles are unable to move blood back up to the heart and lungs. As the blood pools in the legs, swelling and increased pressure cause tiny blood vessels (capillaries) to burst. Red blood cells are released and enter the leg tissue, producing a bluish-purple skin tone. As the condition worsens, the skin surface can be damaged, ulcers develop and infections in the tissue of the leg can result.

DVT symptoms and risk factors

Initial symptoms of DVT can include swelling, pain and warmth in the lower leg, a mild cramping sensation, swollen subcutaneous veins and low-grade fever. The skin may also have a bluish discoloration. However, many DVTs are "silent" and exhibit few or no symptoms. DVT-like symptoms may also be caused by unrelated conditions, including muscle strains and phlebitis (vein inflammation). Careful clinical evaluation is required. Risk factors for DVT include medical conditions such as previous DVT, stroke or heart attack, congestive heart failure, lupus, lower leg fractures and recent major surgery. Extended periods of immobility and long car or plane trips also increase risk.

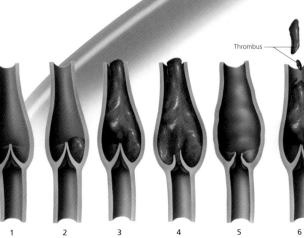

- Thrombus

1   2   3   4   5   6

The formation of blood clots

The series of one-way valves are engineered to allow blood to flow up the leg against gravity. The cross-section illustrations reveal two cusps, which create the two cup-like areas of the valve at risk of pooling blood. The first illustration of the series **(1)** indicates the normal valve closure. The following two illustrations **(2,3)** indicate how pooling can start and then build, creating a clot. The next two illustrations **(4,5)** indicate a complete blockage and irreversible damage to the valve; effective valve closure is compromised and reverse blood flow is possible. A serious condition can result when a section of a blood clot called a thrombus **(6)** breaks off and travels to the lungs, causing an embolism.

Taking control of your DVT

For those who are susceptible to DVT, there are means to limit incidents. Exercising the legs will do the most to reduce the risk of DVT, because it will help keep blood from pooling in the leg veins. If this is not possible, keep the legs elevated when sitting. If the patient is not self-sufficient, aid should be provided to guide the patient's legs through their normal range of motion. A lower leg massage is also beneficial. There are special elastic leg stockings that should be worn. They should be changed once a day to ensure no chafing or leg discoloration has occurred.

Diagnosis and treatment of DVT

When suspicious symptoms are reported, there are several ways to determine if DVT is present. Blood pressure can be checked at various places on the leg, a process called impedance plethysmography (IPG). Also, high-frequency sound waves, or ultrasound, can be used to create images of the leg. X-rays can also be taken of the leg following the injection of a dye-like substance into the veins to highlight blood flow. The standard treatments for DVT typically include blood thinners or anti-coagulants used on a short- or long-term basis.

©Scientific Publishing Ltd., Elk Grove Village, IL. USA
#1652

PLATE 31

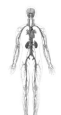

The Effects of Alcohol

The brain
(Sagittal section)

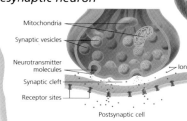

— Striatum
— Nucleus accumbens
— Prefrontal cortex
— Substantia nigra
— Ventral tegmental area

What is alcohol?

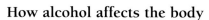

Alcoholic beverages contain **ethanol**, a clear, thin, odorless liquid created by the fermentation of fruit or grain mixtures (wine and beer) or the distillation of fermented fruit or grain mixtures (whiskey, gin, vodka and rum). The exact concentration of ethanol varies according to the type of beverage. On average, beer is 4.5 percent ethanol; wine is 11 percent ethanol; and distilled spirits range from 40 to 95 percent ethanol. Pure alcohol should never be consumed, as it quickly produces effects that can become fatal.

Ethanol
C_2H_6O

How alcohol affects the body

Alcohol affects virtually every part of the body. In the gastrointestinal system, alcohol irritates the linings of the esophagus and stomach, triggers the secretion of acid and histamine, and can cause vomiting. Over time, alcohol use can lead to gastritis or ulcers. Alcohol consumption also increases blood flow to the skin, resulting in lost body heat, while at the same time decreasing blood flow to the muscles. Brain and liver cells are directly affected by alcohol even with occasional drinking.

Long-term effects of heavy drinking include more serious complications, such as:

- Liver enlargement and damage such as alcoholic hepatitis and cirrhosis
- High blood pressure, stroke, irregular heartbeat and heart damage or disease
- Kidney failure resulting from chronic alcohol-induced diuresis
- Increased risk of mouth, larynx, liver and gastrointestinal cancers
- Greater incidence of pneumonia and acute respiratory distress syndrome (ARDS)
- Dietary deficiencies of essential nutrients such as iron, folic acid, and thiamine, which may lead to nerve damage
- Impairment of memory, thinking and concentration skills
- Death of brain cells and reduced brain mass
- Higher risk of injury from falls or accidents
- Decreased production of sex hormones
- Personality changes and other emotional and behavioral problems, including anxiety or depression

Acute overdoses of alcohol, also known as alcohol poisoning, produce symptoms including nausea and vomiting, loss of consciousness, depressed respiration, lack of reflexes, and in severe cases, coma. Blood alcohol concentrations above .40–.50 are considered lethal and may be fatal if not treated immediately.

Nervous system effects

Synaptic knob or axon terminal of presynaptic neuron

Mitochondria
Synaptic vesicles
Neurotransmitter molecules
Synaptic cleft
Receptor sites
Ions
Postsynaptic cell

Dopamine pathways

The dopamine pathway is one component of the brain reward system. This pathway may also be involved in an ethanol reinforcement effect.

Alcohol and neurotransmitters

Alcohol directly affects the function of important chemical messengers in the brain known as neurotransmitters. These highly specialized chemicals stimulate nerve impulses from one neuron to another neuron, muscle or gland, either inhibiting or activating neural impulses. Normal levels of neurotransmitters, such as dopamine, serotonin, opiate neuropeptides, GABA (a major inhibitory transmitter) and glutamate receptors are negatively altered by alcohol consumption.

Cardiovascular effects

Damage to the heart

Cardiomyopathy is a disease of the heart muscle. The muscle fibers are damaged and the heart chamber walls are weakened. To compensate for this injury the chambers of the heart enlarge. The function of the heart is impaired, resulting in inadequate blood flow to the body's organs and tissues. Heart rhythm can be disturbed, with resulting heartbeat irregularities, or arrhythmias. About one-third of cardiomyopathy cases are from excessive alcohol consumption. Alcoholic cardiomyopathy can eventually lead to heart failure.

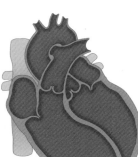
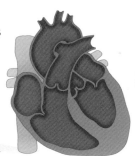

Normal heart

Heart with cardiomyopathy
- *Damaged muscle fibers*
- *Weakened heart chamber walls*
- *Enlarged heart chambers*

Fetal alcohol syndrome

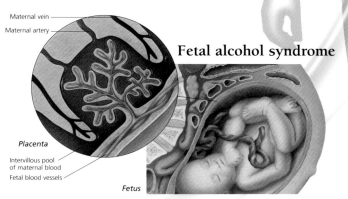

Maternal vein
Maternal artery
Placenta
Intervillous pool of maternal blood
Fetal blood vessels
Fetus

Alcohol in a pregnant woman's bloodstream is passed directly through the placenta and into the developing baby's bloodstream. High alcohol consumption by a pregnant mother can cause fetal alcohol syndrome, which is associated with:

- Mental retardation and developmental delays
- Small body size, slow growth and poor coordination
- Heart defects
- Hearing, vision and dental defects
- Facial abnormalities
- Behavioral problems including hyperactivity and limited attention span

Drinking alcohol during pregnancy also increases the risks of miscarriage, stillbirth and low birth weight.

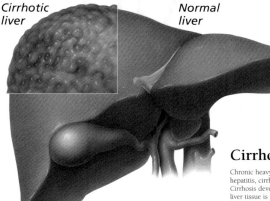

'Area of burst arterioles
Dead brain tissue
Area of damaged arterioles
Circle of Willis

Stroke

Chronic heavy drinking and binge drinking may result in a hemorrhagic and/or ischemic stroke. A hemorrhagic stroke occurs when blood from a ruptured vessel accumulates and compresses surrounding brain tissue, injuring cells and interfering with brain function. The leaking vessel also interrupts oxygen flow to the brain.

Brain in cross section

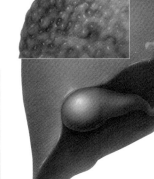

Cirrhotic liver

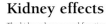

Normal liver

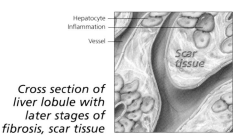

Hepatocyte
Inflammation
Vessel
Scar tissue

Cross section of liver lobule with later stages of fibrosis, scar tissue

Cirrhosis of the liver

Chronic heavy drinking can cause alcoholic hepatitis, cirrhosis or complete liver failure. Cirrhosis develops when a significant portion of liver tissue is progressively and irreversibly destroyed by alcohol abuse.

Kidney effects

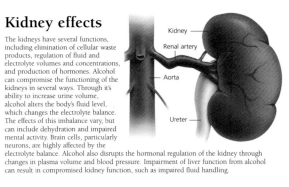

Kidney
Renal artery
Aorta
Ureter

The kidneys have several functions, including elimination of cellular waste products, regulation of fluid and electrolyte volumes and concentrations, and production of hormones. Alcohol can compromise the functioning of the kidneys in several ways. Through its ability to increase urine volume, alcohol alters the body's fluid level, which changes the electrolyte balance. The effects of this imbalance vary, but can include dehydration and impaired mental activity. Brain cells, particularly neurons, are highly affected by the electrolyte balance. Alcohol also disrupts the hormonal regulation of the kidney through changes in plasma volume and blood pressure. Impairment of liver function from alcohol can result in compromised kidney function, such as impaired fluid handling.

Intoxication levels

The degree of alcohol intoxication depends on multiple factors, including body size, the amount and rate of alcohol consumption, the rate of absorption (influenced by the presence or absence of food in the stomach), how the body metabolizes alcohol, genetics and previous drinking experience. As a general rule, alcohol that is consumed slowly (7 grams per hour or approximately 1.5 ounces of 80% proof distilled spirits) will not accumulate in the body or result in intoxication.

BAC*	Effects
.01 – .05 %	Feelings of relaxation, lowered inhibitions
.05 – .07 %	Impairment begins; loss of coordination, reflexes and muscle control; loss of self-control and driving capability
.08 – .10 %	Legally drunk in almost all states and the District of Columbia
.10 – .15 %	Loss of balance, impaired body coordination and slightly slurred speech
.15 – .25 %	Slurred speech, difficulty walking, confusion, loss of perception, vision problems
.25 – .40 %	Most people are in a state of stupor; loss of consciousness, some may die
.40 – .50+ %	Most people are unconscious, breathing shuts off; coma and death are likely

*Approximate Blood Alcohol Concentration

What is alcoholism?

Alcoholism or alcohol dependence is a disease that is usually chronic and progressive and frequently fatal. Symptoms of alcoholism include:

- Emotional and physical dependence on alcohol
- Blackouts and hangovers
- Alcohol-related health problems
- Unpleasant withdrawal symptoms
- Lack of control over the amount or frequency of drinking
- Preoccupation with drinking
- Personality changes and emotional and psychiatric difficulties

Alcohol dependence can take many forms, ranging from occasional drinking to chronic heavy drinking or binge drinking. Most people with alcohol dependence continue to drink even if alcohol is affecting their physical or mental health. Many alcoholics need increasing amounts of alcohol to become intoxicated and experience severe withdrawal symptoms during periods of abstinence. They may also have poor nutrition, gastrointestinal problems, numbness or weakness in the legs and hands, and problems with balance.

Treatment

The first step in treatment is acknowledgement of a drinking problem and a decision to stop drinking. Depending on the severity of the disease, either inpatient or outpatient detoxification may be necessary to help the body reverse its dependence on alcohol. Medical treatment may also include tranquilizers, vitamin supplements and intravenous fluids. Drugs such as disulfiram and naltrexone are sometimes prescribed to reduce the craving for alcohol. Other important components of alcoholism treatment include long-term psychological counseling, self-help groups and counseling for family members.

©Scientific Publishing Ltd., Elk Grove Village, IL, USA
#1653

PLATE 32

The Brain

An average brain weighs between 3 and 3.5 lbs. and is composed of over 100 billion neurons. A brain is divided into 2 structures, the largest being the **cerebrum** (80% of brain mass) and the smaller called the **cerebellum** (20% of brain mass). The cerebrum consists of 2 hemispheres with 5 lobes in each hemisphere. Cerebral hemispheres and lobes each have the specificity of brain function. The cerebral hemispheres control the higher brain functions such as memory, speech and vision, while the cerebellum controls balance and coordination. The brain accounts for about 2% of a person's body weight, yet it receives about 20% of the body's total cardiac output. Interruptions in blood flow to the brain can cause unconsciousness in as little time as 10 seconds or less.

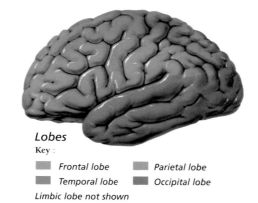

Lobes

Key :

- Frontal lobe
- Temporal lobe
- Parietal lobe
- Occipital lobe

Limbic lobe not shown

Functional areas of the brain

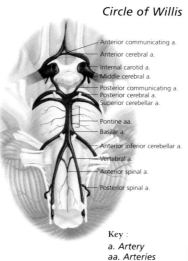

- Primary motor area
- Secondary motor area
- Primary somatosensory area
- Secondary somatosensory area
- Primary visual area
- Secondary visual area
- Primary acoustic area
- Secondary acoustic area
- Sensory speech area

Brain (Base view)

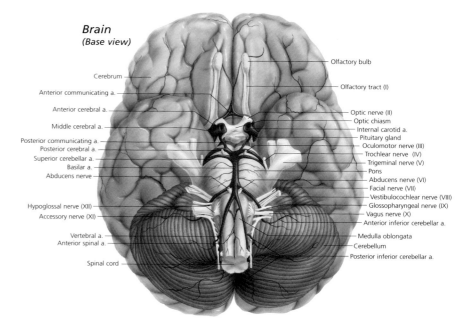

- Cerebrum
- Anterior communicating a.
- Anterior cerebral a.
- Middle cerebral a.
- Posterior communicating a.
- Posterior cerebral a.
- Superior cerebellar a.
- Basilar a.
- Abducens nerve
- Hypoglossal nerve (XII)
- Accessory nerve (XI)
- Vertebral a.
- Anterior spinal a.
- Spinal cord
- Olfactory bulb
- Olfactory tract (I)
- Optic nerve (II)
- Optic chiasm
- Internal carotid a.
- Pituitary gland
- Oculomotor nerve (III)
- Trochlear nerve (IV)
- Trigeminal nerve (V)
- Pons
- Abducens nerve (VI)
- Facial nerve (VII)
- Vestibulocochlear nerve (VIII)
- Glossopharyngeal nerve (IX)
- Vagus nerve (X)
- Anterior inferior cerebellar a.
- Medulla oblongata
- Cerebellum
- Posterior inferior cerebellar a.

Circle of Willis

- Anterior communicating a.
- Anterior cerebral a.
- Internal carotid a.
- Middle cerebral a.
- Posterior communicating a.
- Posterior cerebral a.
- Superior cerebellar a.
- Pontine aa.
- Basilar a.
- Anterior inferior cerebellar a.
- Vertebral a.
- Anterior spinal a.
- Posterior spinal a.

Key :
a. Artery
aa. Arteries

Coronal section

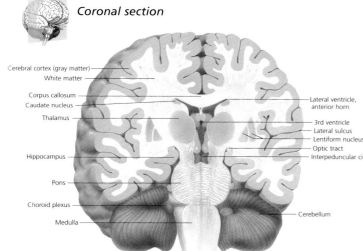

- Cerebral cortex (gray matter)
- White matter
- Corpus callosum
- Caudate nucleus
- Thalamus
- Hippocampus
- Pons
- Choroid plexus
- Medulla
- Lateral ventricle, anterior horn
- 3rd ventricle
- Lateral sulcus
- Lentiform nucleus
- Optic tract
- Interpeduncular cistern
- Cerebellum

Meninges of the brain

- Arachnoid
- Subarachnoid space
- Cerebral vein
- Pia mater
- Scalp
- Periosteum
- Bone
- Epidural space
- Dura mater
- Subdural space
- Arachnoid granulation
- Superior sagittal sinus
- Falx cerebri
- Cerebral hemisphere

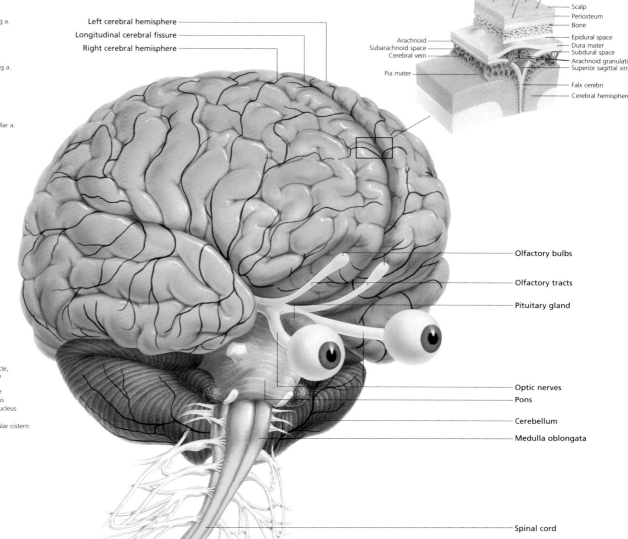

- Left cerebral hemisphere
- Longitudinal cerebral fissure
- Right cerebral hemisphere
- Olfactory bulbs
- Olfactory tracts
- Pituitary gland
- Optic nerves
- Pons
- Cerebellum
- Medulla oblongata
- Spinal cord

Arteries of the brain (Sagittal section)

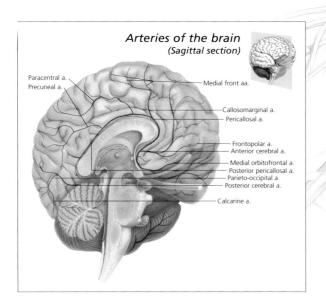

- Paracentral a.
- Precuneal a.
- Medial front aa.
- Callosomarginal a.
- Pericallosal a.
- Frontopolar a.
- Anterior cerebral a.
- Medial orbitofrontal a.
- Posterior pericallosal a.
- Parieto-occipital a.
- Posterior cerebral a.
- Calcarine a.

Limbic system (Sagittal section)

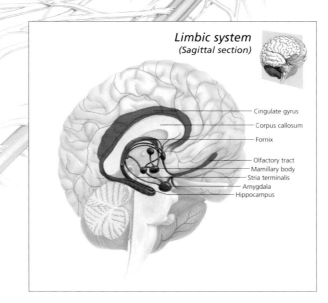

- Cingulate gyrus
- Corpus callosum
- Fornix
- Olfactory tract
- Mamillary body
- Stria terminalis
- Amygdala
- Hippocampus

Ventricles of the brain (Lateral view)

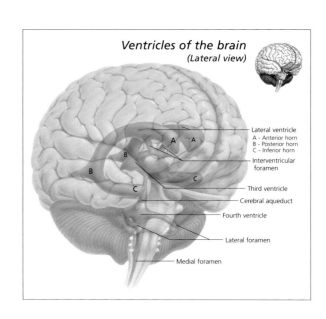

- Lateral ventricle
  - A - Anterior horn
  - B - Posterior horn
  - C - Inferior horn
- Interventricular foramen
- Third ventricle
- Cerebral aqueduct
- Fourth ventricle
- Lateral foramen
- Medial foramen

©Scientific Publishing Ltd., Elk Grove Village, IL. USA
#1700

PLATE 33

Understanding CNS
Central Nervous System

Functional areas of the brain
- Primary motor area
- Secondary motor area
- Primary somatosensory area
- Secondary somatosensory area
- Primary visual area
- Secondary visual area
- Primary acoustic area
- Secondary acoustic area
- Sensory speech area

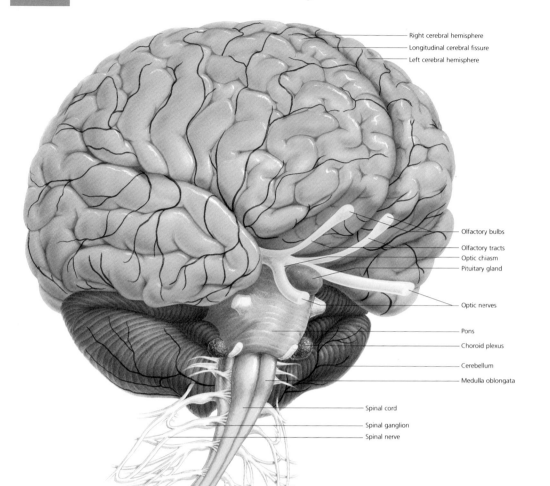

- Right cerebral hemisphere
- Longitudinal cerebral fissure
- Left cerebral hemisphere
- Olfactory bulbs
- Olfactory tracts
- Optic chiasm
- Pituitary gland
- Optic nerves
- Pons
- Choroid plexus
- Cerebellum
- Medulla oblongata
- Spinal cord
- Spinal ganglion
- Spinal nerve

Ventricles of the brain
(CNS circulation)
(Lateral view)

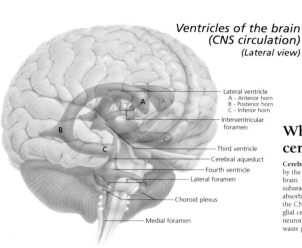

- Lateral ventricle
  - A - Anterior horn
  - B - Posterior horn
  - C - Inferior horn
- Interventricular foramen
- Third ventricle
- Cerebral aqueduct
- Fourth ventricle
- Lateral foramen
- Choroid plexus
- Medial foramen

What is cerebrospinal fluid?

Cerebrospinal fluid (CSF) is mainly secreted by the choroid plexuses of the ventricles of the brain. CSF flows in the ventricles, central canal and subarachnoid space and also acts as a shock absorber. CSF plays many other important roles in the CNS, supplying nutrients to the neurons and glial cells, transporting active biochemicals such as neurotransmitters and hormones, and removing waste products.

What is the central nervous system?

The nervous system is composed of two integrated systems that are responsible for conducting and processing sensory and motor information: the **central nervous system** (CNS) and the **peripheral nervous system** (PNS), which connects the CNS to the rest of the body.

The CNS includes the **brain** and **spinal cord**, which are covered by protective membranes called meninges (dura mater, arachnoid, and pia mater). The brain processes and coordinates all neural signals received from the spinal cord as well as its own nerves, such as the olfactory and optic nerves, and performs complex mental functions such as thinking and learning. The peripheral nervous system transmits input gathered from the sensory organs to the CNS. Motor output signals are relayed back to the PNS and on to the body's muscles and glands. The PNS has three separate divisions called the autonomic, sensory and motor nervous systems.

The functional units of the nervous system are **neurons**. **Sensory neurons** communicate information from sensory receptors to the CNS. **Motor neurons** relay signals from the CNS to effector (muscle and gland) cells. **Interneurons** fill in the spaces between neurons. **Glial cells** also make up a significant portion of the nervous system and provide important support for neuron activity.

Spinal cord

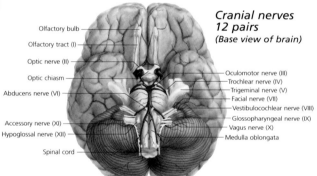

- Central canal
- Anterior fissure
- Gray matter
- White matter
- Posterior (sensory) root of spinal nerve
- Denticulate ligament
- Sensory root ganglion
- Meninges:
  - Pia mater
  - Arachnoid mater
  - Dura mater
- Anterior (motor) root of spinal nerve

Cranial nerves
12 pairs
(Base view of brain)

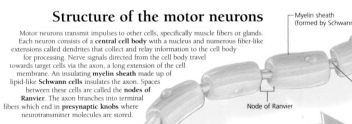

- Olfactory bulb
- Olfactory tract (I)
- Optic nerve (II)
- Optic chiasm
- Abducens nerve (VI)
- Accessory nerve (XI)
- Hypoglossal nerve (XII)
- Oculomotor nerve (III)
- Trochlear nerve (IV)
- Trigeminal nerve (V)
- Facial nerve (VII)
- Vestibulocochlear nerve (VIII)
- Glossopharyngeal nerve (IX)
- Vagus nerve (X)
- Medulla oblongata
- Spinal cord

The spinal cord and nerves

The spinal cord connects the peripheral nervous system to the brain, coordinates simple reflexes to stimuli, and helps regulate the internal organs. It contains 31 pairs of **spinal nerves**, which include both sensory and motor axons. Nerve signals generated by the sensory neurons travel through the spinal cord to the brain. Signals from the motor areas of the brain are sent back through the cord and directed to the motor neurons, triggering a response.

The inner core of the spinal cord is gray matter composed of neuron cells, glial cells, and interneurons. The outer core or white matter is made up of tracts of myelinated axons responsible for transporting nerve signals. Surrounding the spinal cord are the meninges, protective bones of the vertebral column and a cushioning layer of fat and connective tissue in the epidural space.

Synaptic knob or axon terminal of presynaptic neuron

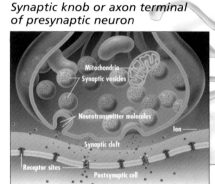

- Mitochondria
- Synaptic vesicles
- Neurotransmitter molecules
- Ion
- Synaptic cleft
- Receptor sites
- Postsynaptic cell
- Axon terminal fiber
- Synaptic knob (or axon terminal of presynaptic neuron)

What are synaptic connections?

Neurons in the CNS create thousands of input and output connections with other neurons, forming dense networks within the brain. Fiber-like structures called **dendrites** extend from the membrane of each neuron to receive and transmit signals from other neurons into the cell body. A long, tube-like extension called the **axon** sends signals from the neuron towards nearby target cells. As impulses arrive at the tip of the axon, the terminal bulbs release "messenger" molecules called **neurotransmitters**. These highly specialized chemicals carry nerve impulses across the tiny space between the axon and the adjacent neurons or cells, either inhibiting or activating neural impulses in the target cell.

Every neurotransmitter (i.e., dopamine and serotonin) has unique characteristics that allow it to bind to specific receptor sites on target cells. However, not all neurotransmitters are absorbed or used by target cells. Neurotransmitter molecules in the synaptic gap may be neutralized by enzyme degradation or reabsorbed back into the axon terminal in a process called **reuptake**, which blocks the neurotransmitter's potential action.

Structure of the motor neurons

Motor neurons transmit impulses to other cells, specifically muscle fibers or glands. Each neuron consists of a **central cell body** with a nucleus and numerous fiber-like extensions called dendrites that collect and relay information to the cell body for processing. Nerve signals directed from the cell body travel towards target cells via the axon, a long extension of the cell membrane. An insulating **myelin sheath** made up of lipid-like **Schwann cells** insulates the axon. Spaces between these cells are called the **nodes of Ranvier**. The axon branches into terminal fibers which end in **presynaptic knobs** where neurotransmitter molecules are stored.

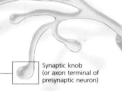

- Myelin sheath (formed by Schwann cells)
- Axon
- Node of Ranvier
- Direction of conduction
- Nucleus
- Nissl bodies
- Cell body
- Dendrites

Neurological disorders

Changes in the normal function of the brain, spinal cord, or the nerves connected to the CNS may be caused by physical trauma, biochemical imbalances, or a variety of other factors. Research and advanced diagnostic techniques are providing new clues into the origin and treatment of many neurological disorders.

Parkinson's disease
- Initiated by degeneration of the area of the brain that produces the neurotransmitter dopamine, which regulates motor activity
- Decreasing supplies of dopamine result in progressive weakness, muscle stiffness, tremors, and difficulty with posture, balance, motion and speech

Depression
- Symptoms may include loss of appetite and interest in activities, sleep disorders, agitation, low energy, and difficulty concentrating
- Linked to abnormal delivery of neurotransmitters, particularly serotonin, acetylcholine, and catecholamines such as adrenaline; also to hormonal imbalances

Epilepsy
- Abnormal electric discharges in the brain and excessive neuron stimulation trigger loss of consciousness and convulsions
- May result from trauma, oxygen deprivation or infection; in isolated cases, from low blood sugar or other imbalances
- Cause is unknown in approximately 50% of cases

Bacterial meningitis
- A dangerous inflammation and infection of the meninges, the protective membranes covering the brain and spinal cord
- Caused by bacteria transported through the blood; may also result from an ear or sinus infection or skull fracture
- Symptoms include high fever, chills, headache, stiff neck, nausea, confusion or coma; immediate treatment is required

Alzheimer's disease
- A degenerative condition that gradually destroys nerve cells in the cerebral cortex, impairing movement, cognition, and memory
- May be influenced by both genetic and environmental risk factors
- Disease process has been linked to impaired flow of nutrients to neurons; reduced levels of the neurotransmitter acetylcholine; and accumulation of an insoluble protein called beta amyloid, which interferes with normal nerve signals and function

©Scientific Publishing Ltd., Elk Grove Village, IL. USA
#1750

PLATE 34

The Liver

The liver weighs approximately 3 lbs. and is the largest internal organ in the body. It is located in the upper right section of the abdomen, behind the rib cage and above the stomach, right kidney and intestines. The liver is divided by connective tissue called the **falciform ligament** into two major lobes. The larger right lobe is approximately six times larger than the smaller left lobe. Two smaller lobes, the quadrate and the caudate, are located on the visceral surface.

The liver performs more than 500 important functions, including its vital role as an entry into the digestive tract and the circulatory system. It is also an important source of **blood storage**.

Approximately 1500 ml of blood flows through the liver per minute. As much as 13 percent of the body's total blood volume is usually contained in the liver, which can swell to hold even larger amounts of blood in response to injury or illness.

The portal system

The hepatic portal system is comprised of a network of veins that transport blood from the internal abdominal organs (stomach, intestine, spleen and pancreas) to the liver. Portal blood contains both **nutrients** and **toxins** that drain directly from the digestive system and must be screened before returning to the body. Specialized macrophages lining the **hepatic sinusoids** called **Kupffer cells** perform this task by detoxifying harmful substances in the blood, destroying old and defective red blood cells, and eliminating bacteria and debris. They also remove nutrients, amino acids and glucose, which are then metabolized by enzymes in the **hepatocytes**, cells that make up most of the liver's structure.

Functions of the liver
The liver is involved in many of the body's metabolic functions, including:
- Production and excretion of bile and cholesterol
- Detoxification of drugs and other harmful substances
- Metabolism of nutrients (fats, carbohydrates and proteins)
- Conversion of excess glucose into glycogen for storage
- Regulation of amino acid levels
- Storage of blood and vitamins including Vitamins A, D, E and K
- Synthesis of plasma proteins and blood clotting factors
- Conversion of ammonia to urea for elimination by the kidneys

■ Venous blood *(filtered)*
■ Portal blood *(unfiltered)*

To Heart
Esophagus
Inferior vena cava
Liver
Hepatic portal vein
Gastric vein
Splenic vein
Spleen
Stomach
Superior mesenteric v.
Right colic vein
Inferior mesenteric vein
Intestinal veins
Appendix

Bile production

A major digestive function of the liver is the production of **bile**, a combination of water, bile salts, bile pigments, phospholipids, cholesterol and other substances. Bile is used during digestion to neutralize stomach acids and emulsify (break down) fats in the **duodenum**, which is attached to the stomach and forms the upper segment of the small intestine. Bile is produced by the **hepatocytes** and secreted into the bile channels for storage in the **gallbladder**, a small, pear-shaped organ on the visceral surface of the liver. Much of the bile used in digestion is **reabsorbed** by the small intestine and later returned to the liver.

Bile secretion is stimulated by a hormone secretin. Bile exits the liver via the right and left hepatic ducts, which join to form a common hepatic duct. This combines with the cystic duct from the gallbladder, creating the common bile duct. Bile travels along the common bile duct to join the pancreatic duct at the duodenum.

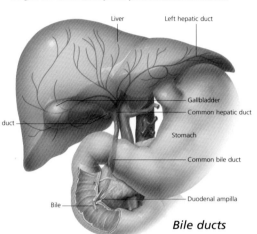

Liver
Left hepatic duct
Gallbladder
Common hepatic duct
Cystic duct
Stomach
Common bile duct
Bile
Duodenal ampilla

Bile ducts

Visceral surface

Fibrous appendix of the liver
Ligamentum venosum
Inferior vena cava
Ligament of the inferior vena cava
Caudate lobe
Left lobe
Right lobe
Portal vein
Hepatic artery
Bile duct
Round ligament of the liver
Quadrate lobe
Gallbladder

Liver lobule function

To the inferior vena cava
Central vein
Hepatic artery
Liver lobule
Branch of bile duct
Branch of portal vein

■ Bile collects in the common bile duct
■ Blood from the digestive system
▢ Oxygenated blood from the heart
■ Deoxygenated and processed blood returning to the heart

The liver possesses the unique capability to regenerate to within 5 to 10% of its original weight after damage from viral or toxic injuries or partial surgical removal. Hepatocytes, stimulated by growth factors after injury, replicate under a process of controlled cell division to restore the liver's volume.

■ Venous blood *(filtered)*
■ Portal blood *(unfiltered)*

Central vein
Sinusoids
Interlobular septum
Central vein
Liver lobule

Portal triad
■ Hepatic artery
■ Branch of portal vein
■ Branch of bile duct

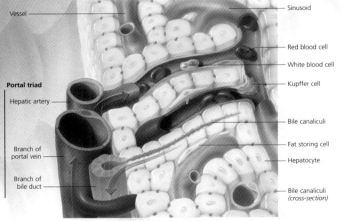

Vessel
Sinusoid
Red blood cell
White blood cell
Kupffer cell
Portal triad
Hepatic artery
Bile canaliculi
Branch of portal vein
Fat storing cell
Hepatocyte
Branch of bile duct
Bile canaliculi *(cross-section)*

Enlarged view of the liver lobule

Structure of the liver

The two lobes of the liver are divided into thousands of functional microscopic units called **lobules**. Blood is delivered to the lobules via the liver's two sources of blood supply: the **hepatic artery**, which carries oxygen-rich blood from the heart, and the **hepatic portal vein**, carrying nutrients from the digestive system.

Each lobule contains layers of **hepatocytes** (liver cells) arranged in cords or sheets radiating out from a central vein. The **hepatic sinusoids** form tunnels or spaces between groups of these layers. Lobules are polygonal in shape, with six **portal triads** at the corners that each contain three vessels: a branch of the hepatic vein, a branch of the hepatic artery and a bile duct. Blood flows into the sinusoids from the portal triads, eventually reaching the **central veins**. Bile flows out of the lobule toward the portal triads through the **bile canaliculi**.

Injury to the liver can result in restricted blood flow through the hepatic portal system, causing portal hypertension, a condition commonly associated with cirrhosis.

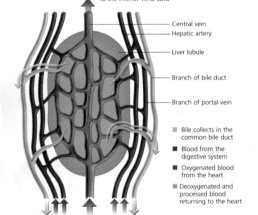

©Scientific Publishing Ltd., Elk Grove Village, IL. USA
#1900
PLATE 35

Understanding Hepatitis

Hepatitis B virus

- Envelope
- Double stranded DNA
- Capsid

Hepatitis viruses vary in their specific characteristics but share a similar structure and infection process. A core of nucleic acid composed of DNA or RNA is surrounded by an outer protein shell with surface proteins or antigens that attach to the host cell. Once attached, viral nucleic acids are released into the host and replicate, generating new virus particles.

- Surface proteins
- Hepatitis B virus

Infection

Infection
is the first stage of hepatitis and occurs when the virus enters the body and invades the liver cells, prompting antibody response. Length of incubation time varies according to the type of virus.

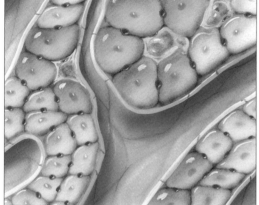

Inflammation
occurs when the immune system responds to infection, causing injury or destruction in the infected liver cells. Inflammation can also result from exposure to drugs, alcohol and other substances.

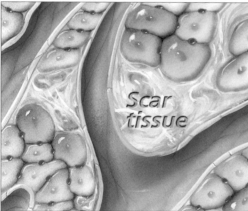

Scar tissue

Fibrosis
is the growth of scar tissue following infection, inflammation or injury to the liver. Over time, scar tissue can inhibit normal liver function, including **blood processing** and **nutrient metabolism**.

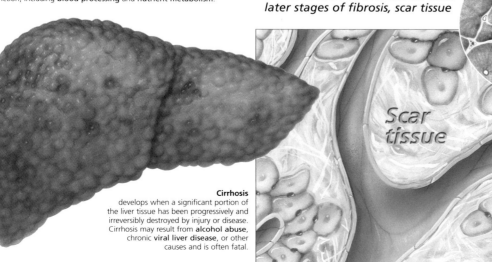

Cirrhosis
develops when a significant portion of the liver tissue has been progressively and irreversibly destroyed by injury or disease. Cirrhosis may result from **alcohol abuse**, chronic **viral liver disease**, or other causes and is often fatal.

What is hepatitis?

Hepatitis is an **inflammation** of the liver triggered by **infection** or **injury** and characterized by the destruction of significant numbers of liver cells. The severity of the disease depends on many factors, including the specific cause of the inflammation as well as any preexisting physical conditions. The symptoms vary widely according to the type of hepatitis and the duration of the inflammation (acute or chronic).

Acute hepatitis
The acute form of hepatitis evolves over a short period of time and resolves **within 6 months** (often 2 months or less). It may be caused by a variety of factors, including hepatitis viruses, medications, toxins and severe bacterial infections.

Chronic hepatitis
Acute hepatitis that persists for **more than 6 months** is called chronic hepatitis. There are many potential causes of chronic hepatitis, including infection by hepatitis viruses B and C. Chronic hepatitis can result in **persistent liver damage**.

Causes of hepatitis

Hepatitis can be caused by any substance or organism that damages the liver, including viruses, autoimmune disorders, alcohol, drugs and chemical toxins.

Viral hepatitis is the leading cause of liver disease in the United States. Symptoms of acute viral hepatitis may begin suddenly or develop over time. They often include nausea and vomiting, slight fever, fatigue, and pressure or pain in the upper right abdomen. Jaundice, dark-colored urine, joint aches, diarrhea and weight loss may also occur.

Autoimmune hepatitis occurs when the body's defense mechanisms attack the liver cells. It can be present in either chronic or acute form and may appear similar to viral hepatitis. Fatigue is the most common symptom, but other symptoms, ranging from mild to severe, can include enlarged liver, jaundice, and joint pain as well as itching and skin rashes. Autoimmune hepatitis is more common in women than men and may be associated with other autoimmune disorders.

Alcohol and drug hepatitis results from excessive or chronic alcohol use or following the consumption of certain drugs or medications. Symptoms include jaundice, fatigue and alterations in sense of taste and smell. **Alcoholic hepatitis** may produce symptoms ranging from mild flu-like characteristics to high fever and enlarged liver. If untreated, this condition can lead to **fatty liver, cirrhosis** or complete **liver failure. Toxic and drug-induced hepatitis** can be caused by a variety of prescription and over-the-counter medications as well as chemical agents and industrial toxins.

Nonalcoholic steatohepatitis can appear in patients with conditions including high cholesterol, diabetes and obesity who do not consume large amounts of alcohol. Symptoms are similar to alcohol-induced hepatitis.

Healthy cross section of liver lobule

Healthy Liver

Liver function

The liver performs more than 500 important functions, including its vital role as a portal between the digestive tract and the circulatory system. It is also an important source of **blood storage**. Approximately 1500 ml of blood flows through the liver per minute. As much as 13 percent of the body's total blood volume is usually contained in the liver, which can swell to hold even larger amounts of blood in response to injury or illness.

What is portal hypertension?

Portal hypertension is an increase in pressure in the portal vein that carries blood from the intestines, spleen and pancreas into the liver.

- In cirrhosis, damage from fibrosis can increase resistance in the portal vein, forcing blood to flow back towards the heart by the way of collateral vessels, instead of through the liver.
- Collateral vessels may develop to bypass the liver, connecting the portal vein directly to the lower portion of the esophagus.
- Swollen collateral vessels (esophageal varices) are fragile and can easily rupture, causing dangerous bleeding in the stomach.

Cross section of liver lobule with later stages of fibrosis, scar tissue

Scar tissue

- Contaminated food or water
- Contact with infected blood
- Alcohol use / Drug abuse
- Sexual contact

Risk factors for hepatitis

- Exchange of bodily fluids with an infected person, especially through sexual contact or sex with multiple partners
- Consumption of contaminated water or food, including improperly cooked shellfish (HAV)
- Contact with infected blood through illicit drug use or occupational needle sticks
- Sharing razors, toothbrushes or other personal items that may contain blood
- Chronic or excessive alcohol use and drug abuse

Enlarged view of the liver lobule

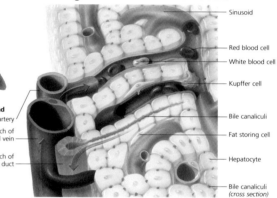

- Sinusoid
- Red blood cell
- White blood cell
- Kupffer cell
- Bile canaliculi
- Fat storing cell
- Hepatocyte
- Bile canaliculi (cross section)

Portal triad
- Hepatic artery
- Branch of portal vein
- Branch of bile duct

Types of hepatitis

There are **six recognized viruses** that can cause viral hepatitis.

HAV
Hepatitis A Virus is an acute form of hepatitis virus that typically begins 2-6 weeks after infection. It is easily spread through **food or water** contaminated by the feces of an infected person or through **contaminated shellfish**. HAV usually requires no treatment and resolves over several weeks, although up to 15% of patients will have prolonged symptoms over a 6-9 month period. HAV can be prevented with **Hepatitis A vaccine**.

HBV
Hepatitis B Virus can cause both acute and chronic hepatitis. Symptoms appear within 1-6 months after infection and may be almost unnoticeable or produce a range of typical viral hepatitis symptoms lasting several weeks or months. HBV is extremely infectious and spreads through **body fluids, contaminated syringes and needles**, or transmission from mother to unborn child. HBV can be prevented with **Hepatitis B vaccine**.

HCV
Hepatitis C Virus causes both acute and chronic hepatitis; it is much more likely to produce **chronic liver disease** than Hepatitis B. Up to 80% of patients have no signs or symptoms of disease, which develop between 2 weeks and 6 months after infection. The chronic form can also develop without early symptoms, although liver damage is occurring. HCV is spread mainly through contact with **infected blood**, including IV drug use, needle sticks, and unprotected sexual contact. There is **no vaccine for HCV**.

HDV
Hepatitis D or Delta Virus occurs only in conjunction with Hepatitis B Virus and spreads through similar routes. Acute infection can be more severe if both viruses are present, although complete recovery usually occurs when infection with both viruses occurs simultaneously (**coinfection**). Patients who develop chronic Hepatitis B and are later infected with HDV (**superinfection**) experience more severe symptoms and are more likely to develop liver failure. HDV can be prevented with **Hepatitis B vaccine**.

HEV
Hepatitis E Virus produces acute hepatitis symptoms and is transmitted via ingestion. It is prevalent in developing countries where **water sources are contaminated** by human waste. Symptoms develop within 2-8 weeks and resolve completely within a month. Pregnant women are at greatest risk from severe illness or acute liver failure. There is **no vaccine for HEV**.

HGV
Hepatitis G Virus is a recently discovered viral form that may occur alone or in the presence of HBV or HCV. Little is known about the course of illness but research suggests that it is mild and short-term. HGV has also been identified in patients with chronic viral hepatitis. It is transmitted via the blood and there is **no vaccine**.

Diagnosis and treatment

Diagnosis is confirmed with clinical tests. Symptoms are also important but may not be present in all patients.
- *Blood tests* measure elevated **bilirubin** and **aminotransferase** levels and the presence of **antigens** and **antibodies** that develop during viral infection. After recovery, antibodies may remain, indicating previous infection. **Immune factors** (serum globulins) in the blood may help diagnose autoimmune hepatitis.
- *Liver biopsy* is required to confirm chronic hepatitis and determine the type and degree of damage from liver disease.
- *Other tests* such as ultrasound and liver/spleen scans may be used to diagnose cirrhosis.

Treatment varies depending on the type and severity of disease.
- Acute infections are typically treated with rest, a balanced diet, and abstinence from alcohol or certain medications that are metabolized in the liver.
- Chronic hepatitis treatment may include interferon and/or other specialized drugs.
- In liver failure, transplantation is the only treatment option.

©Scientific Publishing Ltd., Elk Grove Village, IL. USA
#1950

PLATE 36

The Kidney

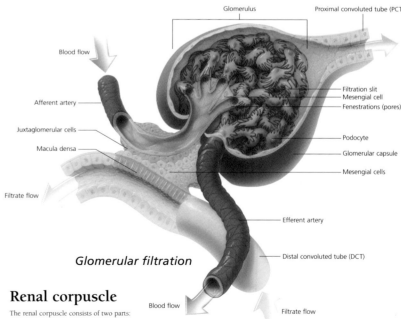

Kidney
Ureter
Bladder

Kidney
Nephron
Renal corpuscle

Segments of the kidney

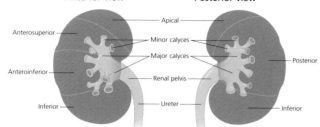

Anterior view

Anterosuperior
Anteroinferior
Inferior

Posterior view

Apical
Minor calyces
Major calyces
Renal pelvis
Ureter

Posterior
Inferior

Structure of the kidneys

The kidneys are located on each side of the spine at the back of the abdominal cavity. Each kidney is approximately 4 to 5 inches long and connects to the bladder via a narrow muscular tube called a **ureter**. We are normally born with a pair of kidneys; however, we can survive with a single kidney.

Each kidney is supported by a layer of connective tissue (**renal fascia**) and surrounded by a **fatty renal capsule**. Within the kidney are two main regions. The outer rim is the **renal cortex**, which contains the **nephrons**, tiny microscopic units that filter blood (*see below*). The inner region is the **renal medulla**. It consists of many cone-shaped structures (**renal pyramids**) that transport urine to the calyces, cup-shaped cavities in the center of the kidney. The calyces drain into the **renal sinus**, a central chamber that connects directly to the ureter.

Functions of the kidneys

The primary function of the kidneys is to filter and eliminate excess water and waste products from the blood. In addition, the kidneys help maintain normal blood pressure by excreting excess sodium and secreting the enzyme **renin**. The kidneys also secrete **erythropoetin**, a hormone essential for the production of red blood cells, and produce active **vitamin D (calcitriol)** to help maintain healthy bones.

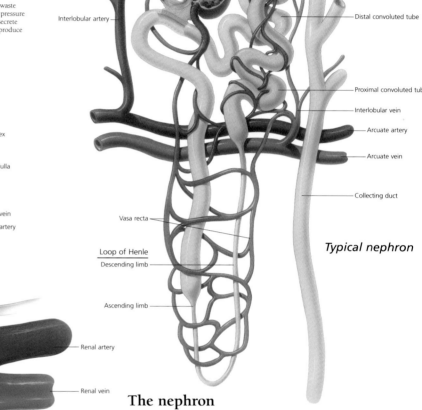

Glomerulus
Glomerular capsule
Interlobular artery

Peritubular capillary
Distal convoluted tube
Proximal convoluted tube
Interlobular vein
Arcuate artery
Arcuate vein

Collecting duct

Vasa recta
Loop of Henle
Descending limb
Ascending limb

Typical nephron

Arcuate artery
Arcuate vein
Renal pyramid
Renal cortex
Renal medulla
Interlobar vein
Interlobar artery

Perirenal fat
Renal fascia

Fibrous capsule
Renal papilla

Major calyx
Minor calyx

Renal artery
Renal vein
Renal pelvis
Fat in renal sinus

Urine
Ureter

The nephron

Each kidney contains more than a million nephrons, the microscopic units located in the outer renal cortex. Nephrons regulate levels of water and soluble substances in the body by filtering the blood, reabsorbing water, glucose, and valuable ions such as potassium and sodium, and excreting excess water and waste products.

A single nephron is made up of four components: the **renal corpuscle** (*see below*), the **proximal convoluted tubule**, the **loop of Henle** and the **distal convoluted tubule**. Blood is first filtered in the renal corpuscle before passing to the proximal tubule, where water and other usable substances are reabsorbed and returned to the bloodstream. The filtrate enters the loop of Henle, and sodium, potassium and chloride are pumped out. In the final stages, additional sodium is removed within the distal tubule and exchanged for potassium and acid. Concentrated urine leaves the nephron via a **collecting duct**.

Filtrate formation

Efferent arteriole
Afferent arteriole

Glomerular capsule
Water & other substances

Collecting tube

Urine
Distal convoluted tubule
Water & other substances

Filtrate flow

Blood flow

How urine is produced

The kidneys process an average of 200 quarts of blood daily, eventually excreting only about 2 quarts of extra water and waste products as urine. Urine production begins when blood enters the nephrons. After a complex process of reabsorption and secretion along the renal tubule, concentrated urine containing water and wastes such as sodium and urea (a byproduct of toxic ammonia products formed in the liver) leaves the collecting ducts. The urine then drains into the calyces of the kidney and enters the ureters, which push small amounts of urine in low pressure waves to the bladder.

Key stages of urine formation

- **Filtration** — filtering of water, waste products, sodium, glucose and other chemicals
- **Reabsorption** — movement of usable substances back to the bloodstream
- **Secretion** — transport of waste materials from capillaries around the renal tubule back into the distal tubule for removal with the urine

Glomerulus
Proximal convoluted tube (PCT)

Blood flow

Afferent artery

Juxtaglomerular cells
Macula densa

Filtrate flow

Filtration slit
Mesangial cell
Fenestrations (pores)
Podocyte
Glomerular capsule
Mesangial cells

Efferent artery
Distal convoluted tube (DCT)

Glomerular filtration

Blood flow
Filtrate flow

Renal corpuscle

The renal corpuscle consists of two parts:

- **Glomerular capsule**, a hollow, cup-shaped bulb that connects to the renal tubule
- The **glomerulus**, a rounded cluster of capillaries where blood cells and larger molecules are filtered

As blood enters the glomerulus at a high pressure, blood is filtered through tiny pores in the glomerular capillaries. The remaining clear fluid containing ions, amino acids, glucose and other substances enters Bowman's capsule, which passes the fluid to the upper part of the renal tubule (PCT) to begin tubular reabsorption.

Adjacent to the renal corpuscle and distal convoluted tubule are specialized cells called the **juxtaglomerular apparatus**. These cells monitor blood pressure as it enters the kidney and react to reduced blood flow by secreting renin, an enzyme stored in the kidney. Renin increases blood pressure by constricting blood vessels and triggering increased reabsorption of sodium and chloride to raise fluid volumes.

©Scientific Publishing Ltd., Elk Grove Village, IL USA
#2000

PLATE 37

The Endocrine System

What is the endocrine system?

The endocrine system is made up of organs and glands that produce **hormones**, internal chemical messengers that regulate and control functions within the body. Hormones are secreted into the bloodstream and trigger activity within a specific organ or tissue by binding to designated receptors to transmit information.

The endocrine system regulates body processes including metabolism and energy balance, reproduction, growth and development, smooth and cardiac muscle contraction, and blood volumes of substances such as sodium and glucose. The activities of the endocrine system are closely coordinated with the nervous system.

The major organs and glands of the endocrine system include the **hypothalamus, thymus, pancreas, ovaries**, and **testes**, as well as the **pituitary, pineal, thyroid, adrenal**, and **parathyroid glands**.

Thymus

The **thymus** produces the thymic hormones, which promote the development of white blood cells called T lymphocytes. The thymus is located between the lungs and behind the breastbone. It is very large at birth but shrinks as thymic tissue is replaced by fat and connective tissue after puberty.

Adult

Juvenile

Heart

Many different endocrine hormones are essential to the function of the cardiovascular system. **Epinephrine** and **norepinephrine** increase heart rate and muscle contractions. **Erythropoietin (EPO)** promotes red blood cell production. **Aldosterone** and **antidiuretic** hormones (ADH) increase the volume of blood.

The endocrine system and hormones

The release of hormones is controlled by **feedback** from different parts of the body. Bursts of hormones are released into the blood in response to signals from the nervous system, as well as by changes in blood chemistry and the actions of other hormones. When sufficient levels of a specific hormone reach the target tissue, stimulation of the hormone-producing organ stops and hormone blood levels decrease.

1. Initial stimulus from hypothalamus

4. Feedback to the pituitary

2. From pituitary through vascular system to target organ

3. Hormonal response

Target organ

Action of hormones

There are many hormones in the circulatory system at the same time. Specific hormones attach to cells having a certain receptor. These cells are called "target" cells. If a cell does not have a receptor, the hormone doesn't connect, and the cells don't respond.

Hormones

Blood vessel

Hormone

Hormone receptor

Target cell

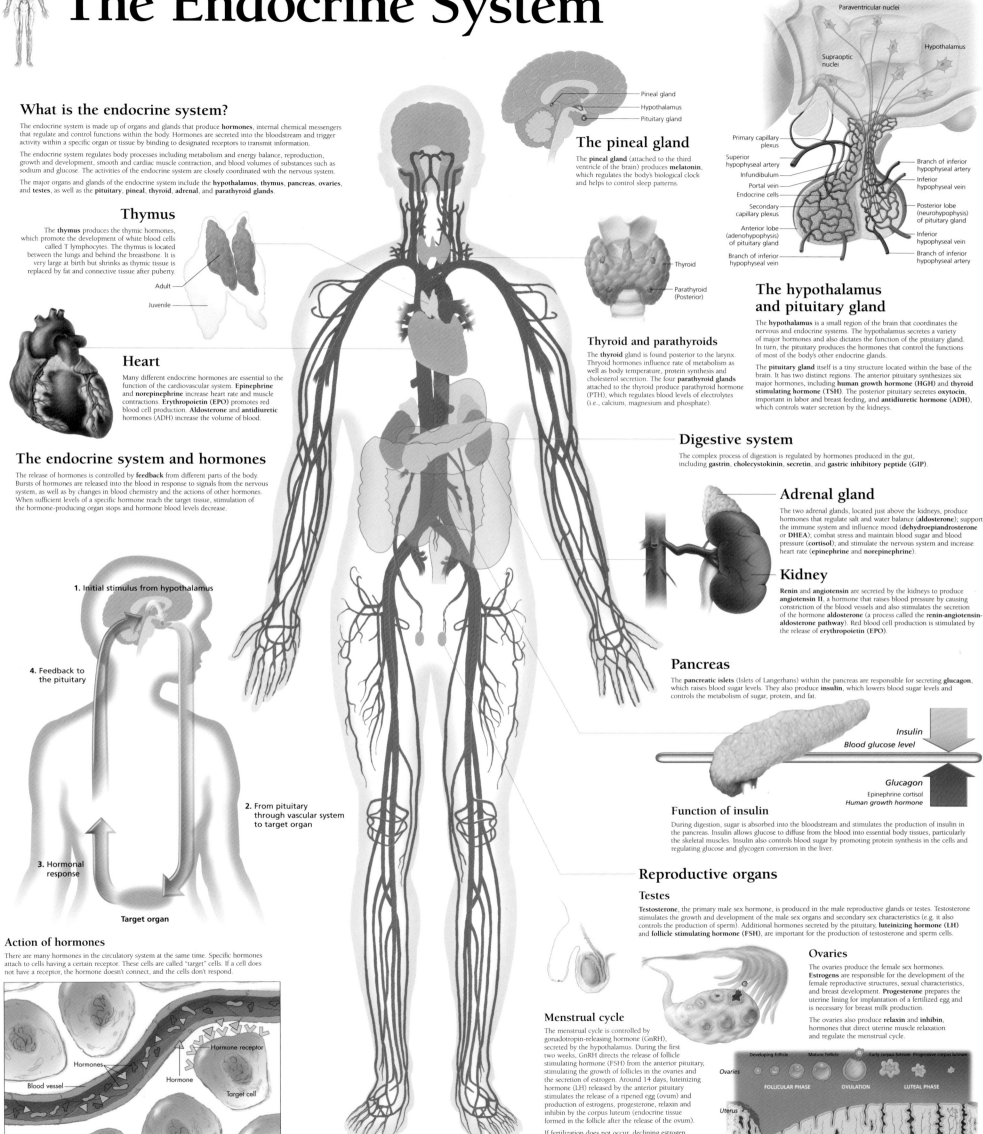

Pineal gland

Hypothalamus

Pituitary gland

The pineal gland

The **pineal gland** (attached to the third ventricle of the brain) produces **melatonin**, which regulates the body's biological clock and helps to control sleep patterns.

Thyroid

Parathyroid (Posterior)

Thyroid and parathyroids

The **thyroid** gland is found posterior to the larynx. Thyroid hormones influence rate of metabolism as well as body temperature, protein synthesis and cholesterol secretion. The four **parathyroid glands** attached to the thyroid produce parathyroid hormone (PTH), which regulates blood levels of electrolytes (i.e., calcium, magnesium and phosphate).

Paraventricular nuclei

Hypothalamus

Supraoptic nuclei

Primary capillary plexus

Superior hypophyseal artery

Infundibulum

Portal vein

Endocrine cells

Secondary capillary plexus

Anterior lobe (adenohypophysis) of pituitary gland

Branch of inferior hypophyseal artery

Branch of inferior hypophyseal artery

Inferior hypophyseal vein

Posterior lobe (neurohypophysis) of pituitary gland

Inferior hypophyseal vein

Branch of inferior hypophyseal artery

Branch of inferior hypophyseal vein

The hypothalamus and pituitary gland

The **hypothalamus** is a small region of the brain that coordinates the nervous and endocrine systems. The hypothalamus secretes a variety of major hormones and also dictates the function of the pituitary gland. In turn, the pituitary produces the hormones that control the functions of most of the body's other endocrine glands.

The **pituitary gland** itself is a tiny structure located within the base of the brain. It has two distinct regions. The anterior pituitary synthesizes six major hormones, including **human growth hormone (HGH)** and **thyroid stimulating hormone (TSH)**. The posterior pituitary secretes **oxytocin**, important in labor and breast feeding, and **antidiuretic hormone (ADH)**, which controls water secretion by the kidneys.

Digestive system

The complex process of digestion is regulated by hormones produced in the gut, including **gastrin, cholecystokinin, secretin**, and **gastric inhibitory peptide (GIP)**.

Adrenal gland

The two adrenal glands, located just above the kidneys, produce hormones that regulate salt and water balance (**aldosterone**); support the immune system and influence mood (**dehydroepiandrosterone or DHEA**); combat stress and maintain blood sugar and blood pressure (**cortisol**); and stimulate the nervous system and increase heart rate (**epinephrine and norepinephrine**).

Kidney

Renin and **angiotensin** are secreted by the kidneys to produce **angiotensin II**, a hormone that raises blood pressure by causing constriction of the blood vessels and also stimulates the secretion of the hormone **aldosterone** (a process called the **renin-angiotensin-aldosterone pathway**). Red blood cell production is stimulated by the release of **erythropoietin (EPO)**.

Pancreas

The **pancreatic islets** (Islets of Langerhans) within the pancreas are responsible for secreting **glucagon**, which raises blood sugar levels. They also produce **insulin**, which lowers blood sugar levels and controls the metabolism of sugar, protein, and fat.

Insulin

Blood glucose level

Glucagon

Epinephrine cortisol
Human growth hormone

Function of insulin

During digestion, sugar is absorbed into the bloodstream and stimulates the production of insulin in the pancreas. Insulin allows glucose to diffuse from the blood into essential body tissues, particularly the skeletal muscles. Insulin also controls blood sugar by promoting protein synthesis in the cells and regulating glucose and glycogen conversion in the liver.

Reproductive organs

Testes

Testosterone, the primary male sex hormone, is produced in the male reproductive glands or testes. Testosterone stimulates the growth and development of the male sex organs and secondary sex characteristics (e.g. it also controls the production of sperm). Additional hormones secreted by the pituitary, **luteinizing hormone (LH)** and **follicle stimulating hormone (FSH)**, are important for the production of testosterone and sperm cells.

Ovaries

The ovaries produce the female sex hormones. **Estrogens** are responsible for the development of the female reproductive structures, sexual characteristics, and breast development. **Progesterone** prepares the uterine lining for implantation of a fertilized egg and is necessary for breast milk production.

The ovaries also produce **relaxin** and **inhibin**, hormones that direct uterine muscle relaxation and regulate the menstrual cycle.

Menstrual cycle

The menstrual cycle is controlled by gonadotropin-releasing hormone (GnRH), secreted by the hypothalamus. During the first two weeks, GnRH directs the release of follicle stimulating hormone (FSH) from the anterior pituitary, stimulating the growth of follicles in the ovaries and the secretion of estrogen. Around 14 days, luteinizing hormone (LH) released by the anterior pituitary stimulates the release of a ripened egg (ovum) and production of estrogens, progesterone, relaxin and inhibin by the corpus luteum (endocrine tissue formed in the follicle after the release of the ovum).

If fertilization does not occur, declining estrogen and progesterone levels decrease blood supply to the lining of the uterus. The lining sloughs off during menstruation, resulting in the menstrual flow.

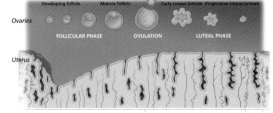

Developing follicle   Mature follicle   Early corpus luteum   Regressive corpus luteum

Ovaries

FOLLICULAR PHASE   OVULATION   LUTEAL PHASE

Uterus

Days 0   4   14   26   0

©Scientific Publishing Ltd., Elk Grove Village, IL. USA
#2800

PLATE 38

The Eye

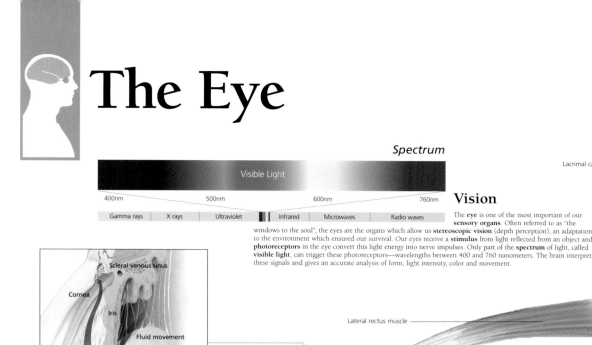

Spectrum

Visible Light			
400nm	500nm	600nm	760nm

| Gamma rays | X rays | Ultraviolet | Infrared | Microwaves | Radio waves |

Vision

The **eye** is one of the most important of our **sensory organs**. Often referred to as "the windows to the soul", the eyes are the organs which allow us **stereoscopic vision** (depth perception), an adaptation to the environment which ensured our survival. Our eyes receive a **stimulus** from light reflected from an object and **photoreceptors** in the eye convert this light energy into nerve impulses. Only part of the **spectrum** of light, called **visible light**, can trigger these photoreceptors—wavelengths between 400 and 760 nanometers. The brain interprets these signals and gives an accurate analysis of form, light intensity, color and movement.

Eyeball · Lacrimal gland · Lacrimal duct · Iris · Sclera (covering) · Pupil · Lacrimal punctum · Lacrimal canaliculi · Lacrimal sac · Nasolacrimal duct

The structure & accessory structures of the eyeball

The wall of the eyeball is made up of three layers. The outermost layer, the **fibrous tunic**, contains both the **sclera** (gives shape to the eyeball) and the **cornea** (transmits and refracts light). The middle layer, the **vascular tunic**, is made up of the **choroid** (supplies blood to the eye), the **ciliary body** (supports the lens and produces a fluid called the **aqueous humor**) and the **iris** (regulates the amount of light by controlling **pupil** size). The innermost layer, the **internal tunic**, contains the **retina** (provides photoreception and transmits impulses). Within the eyeball and suspended from the ciliary body by the **suspensory ligament** is the **lens**, which refracts and focuses the light onto the retina.

Each eye sits in a bony depression of the skull, the **orbit**, which protects and supports the eye while providing a place for attachment for the **extrinsic ocular muscles**, the six muscles that control the movement of the eye. The **eyelids** protect the eye from injury and prevent it from drying out. They distribute the fluid, called tears, that is secreted and drained by the **lacrimal apparatus**. The **eyebrows** and **eyelashes** both protect the eye from airborne and falling particles. The **conjunctiva**, a mucous membrane lining the inside of the eyelids and continuing around the front of the eyeball, prevents objects from sliding around to the back of the eye.

Scleral venous sinus

Scleral venous sinus · Cornea · Iris · Fluid movement · Lens

The **aqueous humor** produced by the **ciliary body** provides nutrients for the lens and cornea, and helps maintain the pressure in the front of the eye. Aqueous humor is reabsorbed back into the bloodstream through the **scleral venous sinus**, located at the junction of the sclera and cornea.

Right eye
(Horizontal section)

Lateral rectus muscle · Scleral venous sinus · Zonular fibers · Iris · Lens · Cornea · Pupil · Aqueous humor · Anterior chamber · Posterior chamber · Ciliary body · Sclera · Ora serrata · Conjunctiva · Choroid · Medial rectus muscle · Vitreous body · Hyaloid canal · Macula lutea · Optic disc · Retinal vessels · Optic nerve · Nerve sheath · Retina

Light Source

Visual field

The **visual field** is the part of the external world that is projected onto the retina. The cornea and lens focus the right part of the visual field onto the left part of the retina of each eye and the left part of the visual field is focused onto the right part of the retina of each eye. Within each eye the visual field is projected upside down and reversed because of refraction.

Information about the visual field travels from the retinas to the brain. Information from the right side of the visual field travels from the left halves of both retinas to the right side of the brain. The signals from the left eye cross the **optic chiasma** to reach the right side of the brain. Information about the left side of the visual field hits the right halves of both retinas and travels to the left side of the brain—the signals from the right eye also cross at the optic chiasma. Within the brain, signals travel to areas responsible for perception and eye and body movements.

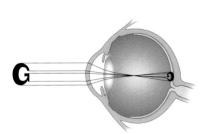

Rods & cones

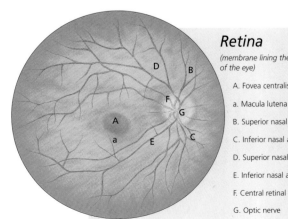

Nerve fibers · Bipolar cells · Rods & Cones · Ganglion cells · Pigment layer

Retina

The **retina**, the internal tunic, is in the posterior part of the eyeball. It is made up of two layers, the **pigmented layer** and the **nervous layer**. The nervous layer is made of three layers of neurons: **ganglion cells**, **bipolar neurons and photoreceptors**. When light hits the retina, it strikes the ganglion cell layer first and then passes through the bipolar layer before reaching the photoreceptors, the **rods** and **cones**. The 120 million rods are the most sensitive to light while the 63 million cones provide color vision and greater visual acuity.

Fovea centralis

The **fovea centralis** is a small pit within the yellow area of the retina called the **macula lutea**. Visual acuity is greatest in the fovea because the area contains only cones and, due to the structure, is the only spot of the retina where light directly hits these photoreceptors.

Optic disc

The **optic disc** is where the nerve fibers from the retina gather together to exit the eye as the **optic nerve**. The disc lacks photoreceptors and is called the eye's blind spot.

Retina
(membrane lining the back of the eye)

A. Fovea centralis
a. Macula lutea
B. Superior nasal artery
C. Inferior nasal artery
D. Superior nasal artery
E. Inferior nasal artery
F. Central retinal artery
G. Optic nerve

Accommodation

The ability of the eye to keep an image focusing on the **retina** is called **accommodation**. When light enters the eye, it is **refracted** or focused onto the retina. In order to keep objects that are moving in focus, the eye has to adjust this refraction. It does this by changing the shape of the **lens** by use of the **ciliary body**. This muscular ring either contracts, making the lens least convex, or relaxes, making the lens more rounded or convex.

Understanding Glaucoma

What is glaucoma?

Glaucoma is a group of diseases in which the normal pressure of the fluid inside the eyes (intraocular pressure) is increased. This can lead to loss of vision and, possibly, to blindness if the condition is left untreated. Glaucoma affects about 1 in 20 people at the age of 70 years. It is more common and may be more severe in black people and in people with a family history of the disease. The most common form of the disease is open angle glaucoma, which affects about two-thirds of glaucoma sufferers.

There is a small space at the front of the eye called the **anterior chamber**. The **ciliary body** supports the lens and produces a watery fluid, the **aqueous humor**, which bathes and nourishes the neighboring tissues. The aqueous humor is drained constantly through a spongy mesh that lies in the angle between the **iris** and the inner surface of the **cornea**. The fluid drains too slowly in people with open angle glaucoma. This causes the intraocular pressure to increase and the cornea to swell. These changes can result in damage to the **optic nerve**. The optic nerve connects the light-sensitive **retina** to the brain and any damage to it can result in defects in vision.

At first, people with glaucoma are free of symptoms. Vision is normal, and there is no pain associated with the condition. As the disease progresses, people may have difficulty moving from a bright room into a darker one and in judging steps and curbs. A person with glaucoma may continue to see objects directly in front of him clearly. However, objects to the side (periphery) may be missed. Blindness can result from a progressive loss of visual field if the disease remains untreated.

The cause of glaucoma

Fluid normally passes through a narrow space between the iris and lens, then drains out of the eye through the **scleral venous sinus**. If this outward flow is blocked, pressure can damage the optic nerve and reduce vision.

How is glaucoma diagnosed?

Glaucoma is a chronic disease that develops slowly. Damage to the optic nerve and visual loss have developed in many patients before the condition has been diagnosed. Regular eye examinations are essential in people who are at risk of developing glaucoma and in those who have been diagnosed with the disease.

A range of tests are used to diagnose and monitor glaucoma:

Visual acuity
The ability of the person to see at various distances is measured using eye charts. These usually consist of letters of different sizes against a plain background.

Tonometry
It has been known for over 100 years that the assessment of intraocular pressure is important in the diagnosis and monitoring of glaucoma. However, it has been recognized more recently that there is no cutoff point between normal and raised intraocular pressure. High intraocular pressure increases a person's risk of developing glaucoma but it does not mean that the person has the disease. The development of glaucoma in an individual depends upon the level of intraocular pressure that the optic nerve will tolerate.

Applanation tonometer

Tonometry is a term used to describe the methods of measuring intraocular pressure. There are many different types of tonometry. The Goldmann applanation tonometer subjects the eye to sufficient force to flatten the cornea. The force applied is measured using a cobalt blue light source and is related to the intraocular pressure by a mathematical equation.

Another type of tonometry relates the time required to flatten the cornea using a puff of air to the intraocular pressure. This method is useful in screening people for glaucoma.

Pupil dilation
Drops are put into the eye to widen the pupil. This allows signs of damage to the optic nerve to be observed more easily.

Visual field
Peripheral vision is measured using the technique of perimetry. The test involves illuminated targets being projected onto an illuminated background. The brightness of the targets is varied so that the average luminescence of the dimmest target can be calculated. The test is repeated several times within the visual field, and any abnormalities are detected using a computer printout.

Perimetry is used to show the extent of any damage due to glaucoma and also to check on any progression of the disease.

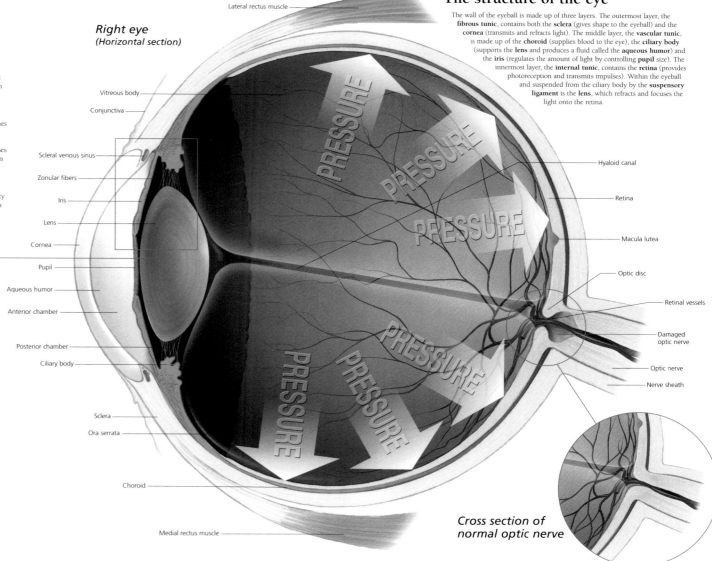

Right eye
(Horizontal section)

Lateral rectus muscle
Vitreous body
Conjunctiva
Scleral venous sinus
Zonular fibers
Iris
Lens
Cornea
Pupil
Aqueous humor
Anterior chamber
Posterior chamber
Ciliary body
Sclera
Ora serrata
Choroid
Medial rectus muscle

Hyaloid canal
Retina
Macula lutea
Optic disc
Retinal vessels
Damaged optic nerve
Optic nerve
Nerve sheath

Scleral venous sinus
Cornea
Iris
Fluid Movement
Lens

The structure of the eye

The wall of the eyeball is made up of three layers. The outermost layer, the **fibrous tunic**, contains both the **sclera** (gives shape to the eyeball) and the **cornea** (transmits and refracts light). The middle layer, the **vascular tunic**, is made up of the **choroid** (supplies blood to the eye), the **ciliary body** (supports the lens and produces a fluid called the **aqueous humor**) and the **iris** (regulates the amount of light by controlling **pupil** size). The innermost layer, the **internal tunic**, contains the **retina** (provides photoreception and transmits impulses). Within the eyeball and suspended from the ciliary body by the **suspensory ligament** is the **lens**, which refracts and focuses the light onto the retina.

Cross section of normal optic nerve

Detail of normal retina

Detail of retina with glaucoma

Area of damaged optic nerve

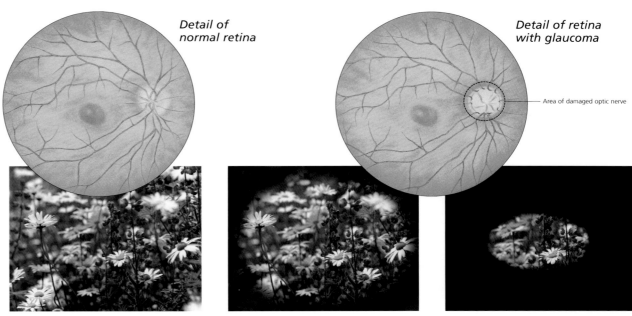

Field of vision loss
The sequence of photographs suggests the progressive narrowing of the field of vision characteristic of glaucoma.

How is glaucoma treated?

The best way to control glaucoma is to ensure that it is detected and treated as early as possible. People who fall into high-risk groups should have their eyes examined regularly. It is essential that people with glaucoma have their condition monitored at regular intervals.

Glaucoma cannot be cured, but it can be corrected in most people. A variety of treatments can be used, depending on the severity of the condition.

Most people with open angle glaucoma are treated with medications. The most common initial treatment is eye drops containing a beta-blocker, such as timolol. These lower the intraocular pressure by reducing the production of aqueous humor and delay the progress of glaucoma. Some people are unable to tolerate beta-blockers or have other medical conditions that prevent their use. Dorzolamide or brinzolamide also reduce the production of aqueous humor and may be suitable for these people. They can also be used in combination with beta-blockers. Other eye drops, such as adrenaline or pilocarpine, can be added as necessary to control intraocular pressure. Apraclonidine is a drug that lowers the intraocular pressure by reducing the rate of production of aqueous humor. Latanoprost lowers the intraocular pressure by increasing the drainage of the aqueous humor from the anterior chamber.

Medication will control intraocular pressure in most people with glaucoma. However, it is essential that medication is used regularly and that glaucoma is monitored to ensure that the combination of drugs selected remains effective.

In some people, medication may not be effective in controlling glaucoma, and surgical treatments have to be considered. In laser surgery, a strong beam of light is focused on the part of the anterior chamber where the aqueous humor is drained from the eye. Changes in the tissue following exposure to the laser result in improved drainage of the aqueous fluid from the eye. The effects of laser surgery may wear off with time. Medication is continued in many patients to maintain long-term benefits.

Surgery is usually reserved for people whose intraocular pressure cannot be controlled by medication or laser surgery. A channel is created in the eye so that the aqueous humor can leave the eye more easily, and intraocular pressure is reduced. Medication is usually required following surgery.

©Scientific Publishing Ltd., Elk Grove Village, IL USA
#2250

PLATE 40

The Ear

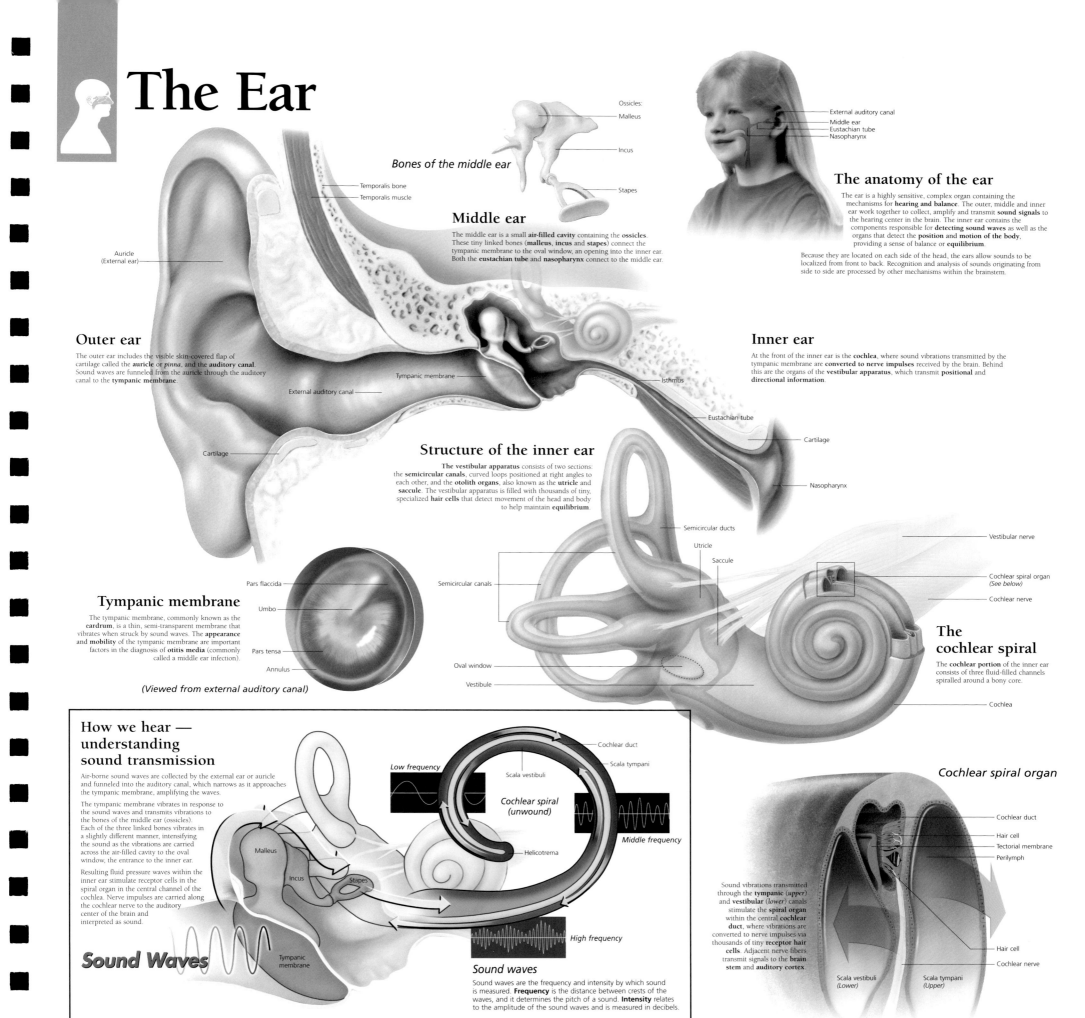

Bones of the middle ear

Ossicles:
- Malleus
- Incus
- Stapes

Middle ear

The middle ear is a small **air-filled cavity** containing the **ossicles**. These tiny linked bones (**malleus**, **incus** and **stapes**) connect the tympanic membrane to the oval window, an opening into the inner ear. Both the **eustachian tube** and **nasopharynx** connect to the middle ear.

- Temporalis bone
- Temporalis muscle

The anatomy of the ear

- External auditory canal
- Middle ear
- Eustachian tube
- Nasopharynx

The ear is a highly sensitive, complex organ containing the mechanisms for **hearing and balance**. The outer, middle and inner ear work together to collect, amplify and transmit **sound signals** to the hearing center in the brain. The inner ear contains the components responsible for **detecting sound waves** as well as the organs that detect the **position and motion of the body**, providing a sense of balance or **equilibrium**.

Because they are located on each side of the head, the ears allow sounds to be localized from front to back. Recognition and analysis of sounds originating from side to side are processed by other mechanisms within the brainstem.

Outer ear

The outer ear includes the visible skin-covered flap of cartilage called the **auricle** or **pinna**, and the **auditory canal**. Sound waves are funneled from the auricle through the auditory canal to the **tympanic membrane**.

- Auricle (External ear)
- Tympanic membrane
- External auditory canal
- Cartilage
- Isthmus
- Eustachian tube
- Cartilage
- Nasopharynx

Inner ear

At the front of the inner ear is the **cochlea**, where sound vibrations transmitted by the tympanic membrane are **converted to nerve impulses** received by the brain. Behind this are the organs of the **vestibular apparatus**, which transmit **positional** and **directional** information.

Structure of the inner ear

The vestibular apparatus consists of two sections: the **semicircular canals**, curved loops positioned at right angles to each other, and the **otolith organs**, also known as the **utricle** and **saccule**. The vestibular apparatus is filled with thousands of tiny, specialized **hair cells** that detect movement of the head and body to help maintain **equilibrium**.

- Semicircular ducts
- Utricle
- Saccule
- Semicircular canals
- Oval window
- Vestibule
- Vestibular nerve
- Cochlear spiral organ (See below)
- Cochlear nerve
- Cochlea

Tympanic membrane

The tympanic membrane, commonly known as the **eardrum**, is a thin, semi-transparent membrane that vibrates when struck by sound waves. The **appearance and mobility** of the tympanic membrane are important factors in the diagnosis of **otitis media** (commonly called a middle ear infection).

- Pars flaccida
- Umbo
- Pars tensa
- Annulus

(Viewed from external auditory canal)

The cochlear spiral

The **cochlear portion** of the inner ear consists of three fluid-filled channels spiralled around a bony core.

How we hear — understanding sound transmission

Air-borne sound waves are collected by the external ear or auricle and funneled into the auditory canal, which narrows as it approaches the tympanic membrane, amplifying the waves.

The tympanic membrane vibrates in response to the sound waves and transmits vibrations to the bones of the middle ear (ossicles). Each of the three linked bones vibrates in a slightly different manner, intensifying the sound as the vibrations are carried across the air-filled cavity to the oval window, the entrance to the inner ear.

Resulting fluid pressure waves within the inner ear stimulate receptor cells in the spiral organ in the central channel of the cochlea. Nerve impulses are carried along the cochlear nerve to the auditory center of the brain and interpreted as sound.

- Low frequency
- Cochlear duct
- Scala tympani
- Scala vestibuli
- Cochlear spiral (unwound)
- Middle frequency
- Helicotrema
- Malleus
- Incus
- Stapes
- Tympanic membrane
- High frequency

Sound Waves

Sound waves

Sound waves are the frequency and intensity by which sound is measured. **Frequency** is the distance between crests of the waves, and it determines the pitch of a sound. **Intensity** relates to the amplitude of the sound waves and is measured in decibels.

Cochlear spiral organ

Sound vibrations transmitted through the **tympanic** (*upper*) and **vestibular** (*lower*) canals stimulate the **spiral organ** within the central **cochlear duct**, where vibrations are converted to nerve impulses via thousands of tiny **receptor hair cells**. Adjacent nerve fibers transmit signals to the **brain stem** and **auditory cortex**.

- Cochlear duct
- Hair cell
- Tectorial membrane
- Perilymph
- Hair cell
- Cochlear nerve
- Scala vestibuli (Lower)
- Scala tympani (Upper)

Macula

The **utricle** and **saccule** each contain a sensory patch called a macula. Tiny hairs in a gelatinous mass move in response to gravity, helping to maintain **equilibrium** by monitoring the position of the head relative to the ground.

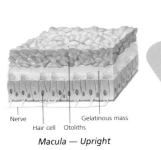

- Nerve
- Hair cell
- Gelatinous mass
- Otoliths

Macula — Upright

- Gravity
- Nerve
- Hair cell
- Gelatinous mass
- Otoliths

Macula — Displaced

Understanding balance

- Semicircular canals
- Utricle
- Saccule
- ● Macula sensors
- ● Crista ampullaris sensors

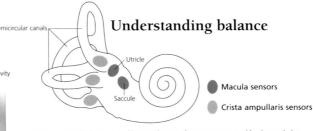

The body's sense of balance or equilibrium relies on information transmitted by the vestibular apparatus, located deep within the inner ear. The **membranous labyrinth** that makes up the vestibular apparatus is filled with **endolymph** fluid, which flows in response to movement of the head and body. The fluid stimulates **tiny hair cells**, triggering sensory neurons that relay information about position and motion to the brain.

Each section within the vestibular apparatus detects and conveys specific nervous impulses. The **semicircular canals** contain receptor structures (**ampulla**) at the base of each canal. Specialized hair cells react to changes in head position, providing **rotational acceleration** information to help the body maintain balance during spinning, tumbling or head-turning motions.

The adjacent **otolith** organs (utricle and saccule) contain sensory structures that monitor **linear acceleration**. By detecting horizontal (back and forth) as well as vertical (up and down) acceleration, these sensory cells help the body gauge how fast it is moving.

Crista ampullaris

The crista ampullaris is in the ampulla at the base of each **semicircular canal**. Sensory hair cells embedded in the cone-shaped gelatinous **cupula** respond to fluid changes in the canal during rotational movement.

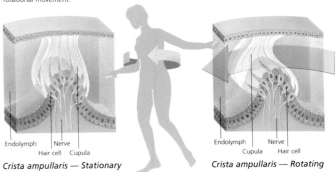

- Endolymph
- Nerve
- Hair cell
- Cupula

Crista ampullaris — Stationary

- Endolymph
- Cupula
- Nerve
- Hair cell

Crista ampullaris — Rotating

©Scientific Publishing Ltd., Elk Grove Village, IL. USA
#2100

PLATE 41

Middle Ear Infections

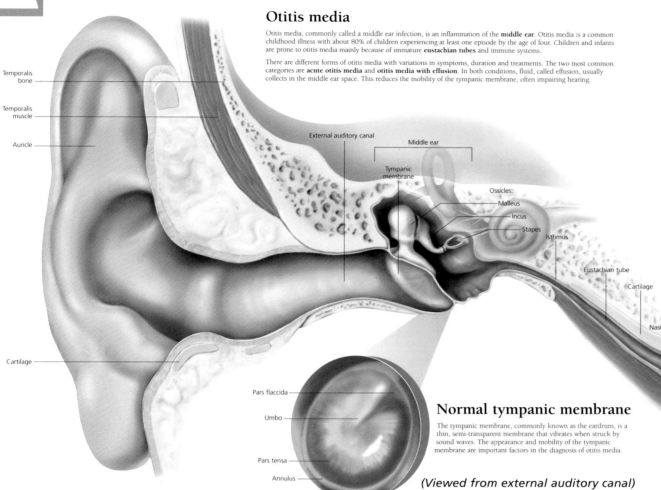

Otitis media

Otitis media, commonly called a middle ear infection, is an inflammation of the **middle ear**. Otitis media is a common childhood illness with about 80% of children experiencing at least one episode by the age of four. Children and infants are prone to otitis media mainly because of immature **eustachian tubes** and immune systems.

There are different forms of otitis media with variations in symptoms, duration and treatments. The two most common categories are **acute otitis media** and **otitis media with effusion**. In both conditions, fluid, called effusion, usually collects in the middle ear space. This reduces the mobility of the tympanic membrane, often impairing hearing.

Temporalis bone
Temporalis muscle
Auricle
Cartilage
External auditory canal
Middle ear
Tympanic membrane
Ossicles:
Malleus
Incus
Stapes
Isthmus
Eustachian tube
Cartilage
Nasopharynx

Pars flaccida
Umbo
Pars tensa
Annulus

Normal tympanic membrane

The tympanic membrane, commonly known as the eardrum, is a thin, semi-transparent membrane that vibrates when struck by sound waves. The appearance and mobility of the tympanic membrane are important factors in the diagnosis of otitis media.

(Viewed from external auditory canal)

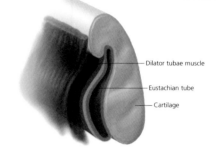

External auditory canal
Middle ear
Eustachian tube
Nasopharynx

Middle ear

The middle ear is a small, air-filled cavity located behind the tympanic membrane that contains tiny bones called ossicles. The middle ear is connected to the respiratory system by the eustachian tube. The mucosal lining of the respiratory system is continuous with the lining of the eustachian tube and middle ear. This connection makes the middle ear susceptible to respiratory system inflammation and infection.

Eustachian tube in cross-section

Dilator tubae muscle
Eustachian tube
Cartilage

Eustachian tube

The eustachian tube normally lies flat but is opened by the dilator tubae muscle during yawning, sneezing and swallowing. When functioning properly, the eustachian tube performs three main functions:

- Ventilation of the middle ear by intermittent opening. This keeps middle-ear air pressure equal to outside air pressure.
- Drainage of middle ear secretions into the nasopharynx.
- Protection of the middle ear from nasopharyngeal secretions and sounds.

Acute otitis media

Acute otitis media is a painful middle ear inflammation in which effusion collects in the middle ear. It is characterized by a rapid onset of the symptoms of acute infection. The infection is usually caused by bacteria, although it can be viral.

Acute otitis media often follows an upper respiratory tract infection. Bacteria in the middle ear can proliferate in the effusion or can be aspirated into the middle ear from the nasopharynx.

If middle ear effusion persists for more than three months after an episode of acute otitis media, the condition becomes **chronic otitis media with effusion**.

Symptoms

Earache	Nausea and vomiting
Fever	Diarrhea
Hearing loss	Loss of appetite
Discharge from ear	Irritability

Bacteria

Streptococcus pneumoniae is the most common bacteria associated with acute otitis media.

Tympanic membrane

- Thickens and becomes opaque
- Appears full or bulging
- Lacks mobility

Treatments

The treatment of otitis media varies according to the type, severity and duration of the condition. Sometimes active treatments are not necessary, and watchful waiting is the most appropriate course of action.

Antibiotics

Your doctor will often prescribe a course of antibiotics to eliminate any bacterial infection in the middle ear or respiratory system. There is currently a new series of vaccines for preventing the most common causes of middle ear infections.

Surgical management

Myringotomy and tympanostomy tube insertion can be used to ventilate and drain an effusion in the middle ear. The myringotomy can be performed as a separate procedure or in conjunction with the insertion of a tympanostomy tube.

Myringotomy
An incision is made in the tympanic membrane.

Tympanostomy Tube
A tympanostomy tube is inserted into the incision.

Eustachian tube dysfunction

Improper eustachian tube functioning is a common factor causing otitis media. Because children's eustachian tubes are not structurally or functionally mature, they are especially prone to these problems. Children's eustachian tubes are shorter and more horizontal than adults', allowing secretions from the nasopharynx to pass more easily into the middle ear. Due in part to having more flexible cartilage, the opening and closing mechanism of a child's eustachian tube often does not function properly.

Obstruction

Obstruction of the eustachian tube prevents middle ear secretions from draining into the nasopharynx, causing fluid to accumulate in the middle ear. The tube can become blocked because of allergies or enlarged tonsils, but often the culprit is an upper respiratory tract infection. The respiratory mucosa becomes congested, and the narrowest part of the eustachian tube, the **isthmus**, becomes blocked.

Aspiration

Secretions in the nasopharynx, often containing bacteria, can be aspirated up the eustachian tube and into the middle ear. Two reasons that aspiration of fluid can occur:

- The eustachian tube stays open, allowing secretions to flow into the middle ear. Lying horizontally makes it easier for the fluid to reach the middle ear.
- The eustachian tube does not open properly, so air pressure in the middle ear cannot be regulated. A negative pressure is created within the middle ear, which pulls the fluid from the nasopharynx up the eustachian tube.

Middle ear
Isthmus of eustachian tube
Nasopharynx
Middle ear
Nasopharynx
SECRETIONS

Otitis media with effusion

Otitis media with effusion is an inflammatory condition in which fluid collects in the middle ear, but there are no symptoms of acute infection. Otitis media can occur in conjunction with an upper respiratory tract infection, after an episode of acute otitis media, or independent of other illnesses. Bacteria may or may not be present in the effusion.

Otitis media with effusion differs from acute otitis media in that it is generally a painless condition. For this reason, otitis media with effusion may be present for a long time before being diagnosed.

Otitis media with effusion often resolves spontaneously without treatment. If the effusion fails to resolve after a reasonable amount of time, surgical management may be indicated.

Symptoms

Otitis media with effusion can be asymptomatic, but if symptoms are present, they are often vague. Mild to moderate hearing loss and a feeling of fullness or ringing in the ear are most common.

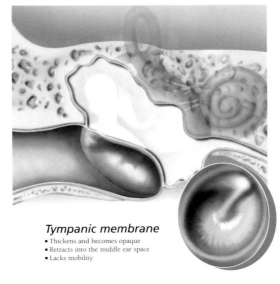

Tympanic membrane

- Thickens and becomes opaque
- Retracts into the middle ear space
- Lacks mobility

Ruptured tympanic membrane

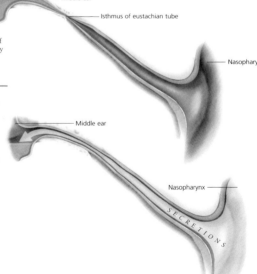

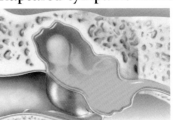

As a result of pressure effusion from acute otitis media in the middle ear, the tympanic membrane can spontaneously rupture. This opening allows drainage of fluid or pus from the middle ear into the external auditory canal and eustachian tube. The perforation generally closes within 3 weeks.

©Scientific Publishing Ltd., Elk Grove Village, IL USA
#2150

PLATE 42

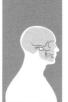

Understanding the Teeth

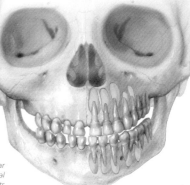

The anatomy of the teeth

The teeth are living, calcified structures embedded in the upper (**maxillary**) and lower (**mandibular**) arches of the jaw. The part of the tooth visible above the gumline is called the **crown**. Below the gumline is the **root**, which extends into the bony portion of the jaw. The teeth are tightly surrounded by soft tissue called the **gingiva** (gums) and cushioned by shock-resistant **periodontal membrane**, which lines the bony sockets within the jaw.

Deciduous dentition

Deciduous (baby) teeth are the first, temporary set of teeth. Beginning with the lower incisors, deciduous teeth typically erupt between the ages of 6 and 24 months. There are 20 deciduous teeth (10 upper and 10 lower), which remain in place until they are shed and replaced by the permanent (adult) teeth beginning around age 6, during a process known as **exfoliation**. By age 13, baby teeth are usually completely replaced by permanent teeth.

Healthy baby teeth play a key role in a child's ability to form clear speech, chew efficiently and develop normal jaw structure and facial characteristics. Extensive decay or tooth loss can have lasting effects on the appearance and development of the child's permanent teeth.

■ Permanent teeth

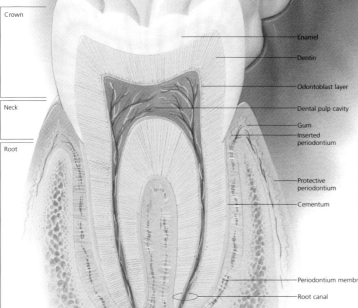

Crown

Neck

Root

- Enamel
- Dentin
- Odontoblast layer
- Dental pulp cavity
- Gum
- Inserted periodontium
- Protective periodontium
- Cementum
- Periodontium membrane
- Root canal
- Apical foramen
- Root apex
- Mandible
- Artery, vein and nerves

A tooth consists of four layers

Enamel is the white, highly calcified outer layer. It is the hardest substance in the body and highly resistant to acids and other corrosive agents. Enamel is without feeling.

Dentin is a hard, yellow layer of tissue beneath the enamel that forms the bulk of the crown. It is softer than enamel and transmits sensations such as temperature and pain to the root.

Cementum is a thin, bony layer covering the root portion of the tooth. It is connected to the jaw bone by collagen fibers that pass through the periodontal membrane to hold the tooth in place.

Pulp is the soft tissue in the inner cavity of the tooth. It contains the nerve fibers and blood vessels and supplies nutrients to the tooth. Pulp extends into the jaw bone and is highly sensitive to pain and temperature.

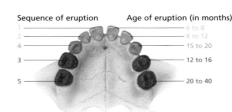

Sequence of eruption | Age of eruption (in months)

1	6 to 8
2	8 to 12
4	15 to 20
3	12 to 16
5	20 to 40

Sequence of eruption | Age of eruption (in years)

2	6 to 9
1	7 to 10
5	10 to 14
4	9 to 13
6	10 to 14
8	17 to 30

Child — Deciduous dentition

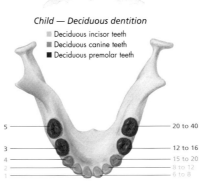

■ Deciduous incisor teeth
■ Deciduous canine teeth
■ Deciduous premolar teeth

Adult — Permanent dentition

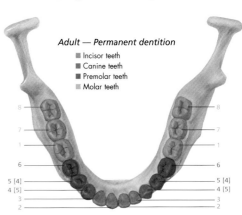

■ Incisor teeth
■ Canine teeth
■ Premolar teeth
■ Molar teeth

5	20 to 40
3	12 to 16
4	15 to 20
2	8 to 12
1	6 to 8

8	8
7	7
1	1
6	6
5 [4]	5 [4]
4 [5]	4 [5]
3	3
2	2

Types and functions of teeth

The adult jaw holds **32 permanent teeth** arranged in an arch, with 16 teeth on the upper jaw and 16 teeth on the lower. The general positions of the teeth within the mouth are noted as either anterior (towards the front) or posterior (towards the back). There are four types of permanent teeth:

ANTERIOR

Incisors
- Sharp, chiseled shape
- Located at the center front of the mouth
- Used to cut or shear food
- 2 central upper/lower, 2 lateral upper/lower

Canines
- Also called cuspids
- Shaped like points
- Work with the incisors to tear food
- Support the lips and guide jaw alignment
- 2 upper, 2 lower

POSTERIOR

Premolars
- Also called bicuspids
- Broad surfaces with pointed cusps
- Used to crush and tear food
- Support vertical dimension of the jaw and face
- 4 upper, 4 lower

Molars
- Broad surfaces with several cusps
- Important for grinding food
- Work with premolars to maintain vertical dimension
- 6 upper, 6 lower

❶ Enamel decay
❷ Dentin decay
❸ Pulp inflammation
❹ Death of the pulp
❺ Abscess

Tooth decay and treatment

Dental caries, also known as tooth decay (cavities), is primarily caused by the deposit of **plaque** on the surfaces of the teeth. Plaque consists of a thin, sticky film of **food debris**, **mucus** and **bacteria**. Although natural protective bacteria in the saliva help protect the teeth by neutralizing acids and washing away food particles, bacterial activity in the plaque can still form erosive acids that gradually destroy the enamel of the tooth. Once a cavity begins, erosion continues into the dentin and if left untreated, to the pulp and nerve of the tooth.

Treatment

Because early tooth decay does not cause pain, **regular dental checkups** are important to detect caries before the decay spreads. Once a cavity is identified, the decayed portion of the tooth is restored with a filling and a crown, if necessary. Tooth decay that has reached the pulp will require additional treatment, such as antibiotics and root canal. *(See also periodontal disease.)*

Root canal procedure

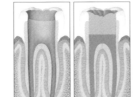

- Inflamed or infected pulp is removed
- Tooth and root canal are cleaned, filled and sealed
- Crown is fitted to protect tooth

Filling

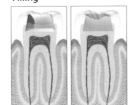

- Bacteria and debris in cavity are cleaned
- Silver/amalgam or composite filling is applied

(filling depends on location and extent of cavity)

Why do teeth hurt?

Pain either in one tooth or in the teeth and gums generally can occur as a sharp twinge or a dull throb.

Tooth pain during or after eating or in response to hot or cold temperatures may be caused by tooth decay, gingivitis or linked to recent dental work

Repeated episodes of throbbing pain are usually associated with advanced tooth decay/inflammation of the pulp (pulpitis)

Continuous pain and/or elevated temperature result from severe pulpitis or an abscess in the pulp or root canal

Impaction

Teeth become impacted when they remain embedded in the gum (**gingiva**) or bone and either fail to emerge or emerge only partially. Impaction occurs when a tooth is blocked by other teeth, because the jaw is too small or if teeth become tilted or twisted as they emerge. The most common type of impaction is in the wisdom teeth, or third molars. Symptoms may include pain in the gum or jaw, inflammation in the gum around the tooth caused by trapped debris (**pericoronitis**) and prolonged headache or jaw ache. Pressure from impacted teeth may also cause misalignment in nearby teeth. Extraction is usually recommended for symptomatic impactions.

Horizontal

Mesioangular

Periodontal disease

Periodontitis is an inflammation of the periodontal ligaments, gingiva, cementum and bone. It is the leading cause of tooth loss in adults and usually occurs as a result of untreated **gingivitis** (infection and inflammation of the gums). Pockets of plaque and tartar (**calculus**) develop around the base of the teeth and become trapped, resulting in continued inflammation that eventually destroys bone and gum tissue holding the teeth in place. Symptoms include swollen, bright red and/or shiny gums that bleed easily, gum tenderness or pain, and loose teeth. Treatment involves extensive cleaning to remove deposits (**scaling**) and/or surgery to support weakened teeth or remove damaged teeth.

Normal | Gingivitis | Periodontitis | Periodontal disease – Moderate | Periodontal disease – Severe

Occlusion

The alignment of the upper and lower jaws and surfaces of the teeth is called occlusion (or **bite**). In many people, occlusion abnormalities (**malocclusion**) occur as a result of disproportionate teeth and jaw size, extra teeth, tooth loss, trauma and other factors. Surgery and/or **orthodontic** treatment may be necessary to reposition and align the teeth.

Types of malocclusion Type 1: **Overcrowding** or poor positioning of the teeth. Type 2: **Underbite** (protruding lower jaw and teeth). Type 3: **Overbite** (upper jaw overlapping the lower).

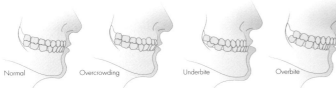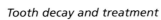

Normal | Overcrowding | Underbite | Overbite

©Scientific Publishing Ltd., Elk Grove Village, IL. USA
#2300

PLATE 43

The Lymphatic System

What is the lymphatic system?

The **lymphatic system** is an extensive network of vessels and nodes that forms a central part of the body's defenses against illness and injury. Foreign materials such as bacteria or dead cells are collected and transported through the **lymph vessels**, where they are filtered by the **lymph nodes**. The lymph vessels also drain excess fluid from the body's tissues, forming a fluid called **lymph**, and carry substances such as cholesterol and fat-soluble vitamins from the gastrointestinal system to the bloodstream.

The lymphatic system contains millions of **lymphocytes**, cells that trigger immune responses, target and destroy pathogens, and produce **antibodies**, proteins that inactivate specific antigens. The lymphatic system is closely integrated with two kinds of lymphatic organs and tissues throughout the body.

Primary lymphatic organs. The **bone marrow** produces cells that divide and mature to become B lymphocytes, or migrate to the **thymus** to develop into T lymphocytes. Each type of lymphocyte plays a specific role in immune response.

Secondary lymphatic organs. Most immune responses actually take place in the secondary lymphatic organs and tissues such as the **lymph nodes**, **spleen**, **tonsils** and **mucosa-associated lymphatic tissue (MALT)**, located throughout the linings of the gastrointestinal, urinary, reproductive and respiratory systems.

Right lymphatic drainage

Left lymphatic drainage

Venous capillary
Lymphatic capillary
Arterial capillary
Excess fluid
Excess fluid

Right lymphatic duct — Left lymphatic duct
Thymus
Axillary nodes
Thoracic duct
Cisterna chyli
Spleen
Lumbar nodes
Inguinal nodes
Popliteal nodes
Lymphatics

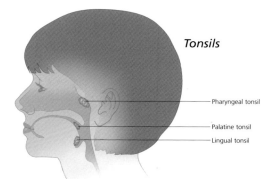

Tonsils

Pharyngeal tonsil
Palatine tonsil
Lingual tonsil

Lymph and lymph vessels

Lymph vessels begin as microscopic capillaries in intercellular spaces throughout the body and converge to form larger lymphatic vessels similar to veins. The lymph capillaries absorb excess fluid filtered from the blood to provide the tissues with oxygen and nutrients. Once inside the lymph vessels, this clear, watery **interstitial fluid** becomes known as **lymph**. While most interstitial fluid is reabsorbed by the blood vessels, approximately 15% returns to the blood system through the lymph vessels. This function helps to maintain the fluid volume of the body.

All lymph passing through the lymphatic system is filtered by the lymph nodes lining the vessels. The lymph eventually flows into large channels called the **thoracic** and **right lymphatic ducts** and drains back into the bloodstream through the **subclavian veins**.

Lymph node

Afferent lymphatic vessels
Lymph flow
Germinal center
Cortical nodule
Capsule
Hilum
Valve
Trabecula
Efferent lymphatic vessel
Vein
Artery
Lymph flow

Lymph nodes

The lymphatic vessels are lined with hundreds of tiny bean-shaped organs called **lymph nodes**. Although they are scattered throughout the body, large clusters of lymph nodes are concentrated near specific areas such as the **mammary glands** and **groin**. Lymph nodes act as a barrier to infection by scavenging bacteria and other foreign materials from the lymph collected from the organs and tissues before it is returned to the bloodstream.

Each lymph node is surrounded by a thick capsule of fibrous tissue and densely packed with lymphocytes, including T cells, antibody-secreting B cells, and scavenging cells called **macrophages**, which ingest foreign matter and debris trapped by specialized fibers within the node's lymphatic sinuses.

Swollen lymph nodes often indicate the presence of infection or cancer.

Lymphoid intestinal tissue

Groups of lymphatic nodules known as **mucosa-associated lymphatic tissue (MALT)** are located throughout the mucosal linings of many areas of the body. In the gastrointestinal tract, concentrations of lymphatic nodules are referred to as **gut-associated lymphoid tissue (GALT)**. This important secondary lymphatic tissue includes large specialized aggregates called **Peyer's patches**, and is believed to play an important role in defending the gut from harmful bacteria and other foreign substances.

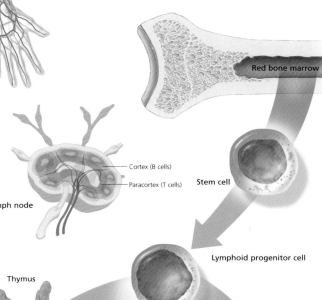

Red bone marrow
Cortex (B cells)
Paracortex (T cells)
Lymph node
Stem cell
Lymphoid progenitor cell
Thymus
NK cell
B cell
T cell

Inflammatory and immune responses

There are two basic kinds of immunity. **Innate** or **nonspecific immunity** is present at birth and involves white blood cells including **monocytes**, **neutrophils** and **eosinophils**, and certain lymphocytes called **NK (natural killer) cells**. These cells play similar roles in attacking and destroying bacteria and other foreign substances. Inflammation is a nonspecific response to invasion by pathogens, injury such as abrasions or cell disturbances, and severe temperature.

Through **specific** or **adaptive immunity**, the body learns and remembers the responses needed to destroy specific antigens. Lymphocytes play a key role in specific immune responses because they are long-lived cells that "remember" the specific antigens they encounter and can react quickly and directly each time the antigen enters the body. Specific immunity involves two different types of responses:

■ **Cell-mediated responses.** This process triggers killer, or cytotoxic, T lymphocytes to directly attack the antigen and is typically targeted against pathogens that invade host cells, such as viruses and some cancer cells.

■ **Antibody-mediated responses.** Antibodies synthesized from **B cells** are produced to inactivate the attacking antigen. This response works primarily against pathogens such as bacteria that multiply in body fluids without entering the body.

Complement proteins. The antimicrobial activity of complement proteins enhances or complements the body's immune, allergic and inflammatory reactions. Complement proteins are found in the blood plasma and on the plasma membranes.

Immune system disorders

Immune disorders occur when the immune system mistakenly reacts against itself and destroys healthy cells and tissues (autoimmune disorders and allergic reactions), or fails to generate sufficient immune response to protect the body from invading pathogens (immunodeficiency).

Autoimmune disorders can occur as a result of injuries, viruses, radiation, or by ingesting certain drugs or other foreign substances. They can also result from a malfunction of the B lymphocytes, which produce abnormal antibodies. Autoimmune disorders include **Type 1 diabetes, lupus, pernicious anemia** and **Graves' disease.**

Allergic reactions involve an excessive immune reaction to a normally harmless substance. Allergic reactions can be caused by exposure to allergens ranging from **pollen** and **dust** to **drugs, food** and **animal dander.** Most allergic symptoms are mild, but some can trigger asthma or sudden anaphylactic reactions requiring immediate treatment.

Immunodeficiency disorders impair the body's ability to defend itself against pathogens, resulting in abnormally frequent or severe infections. Immunodeficiency disorders may be congenital or develop as a result of a secondary disease or infection, such as **human immunodeficiency virus (HIV).** Immunodeficiency can also be caused by **malnutrition** and **immunosuppressant drugs.**

Formation of lymphocytes

All lymphocytes originate as **stem cells** during fetal development. Specific lymphocyte growth and production takes place in the primary lymphatic organs (bone marrow and thymus gland), where the stem cells divide and mature into B and T cells.

B lymphocytes remain in the bone marrow and develop unique receptors targeted to specific antigens. B cells secrete antibodies, large protein molecules that combine with and destroy invading antigens. B cells ultimately form millions of different kinds of antibodies, each with specific antigen-targeting properties.

T lymphocytes begin in the bone marrow and migrate to the thymus gland, located between the lungs and behind the breastbone. They divide rapidly, eventually developing reactive properties against millions of different antigens. Through a complex process within the thymus, T cells become "self-recognizing" so they react only against outside antigens, not against other lymphocytes. Killer, or cytotoxic, T lymphocytes recognize and destroy infected or abnormal cells.

NK (natural killer) cells are lymphocytes related to T cells. NK cells require no previous activation, providing a nonspecific response to virus-infected and malignant cells.

©Scientific Publishing Ltd., Elk Grove Village, IL. USA
#2400

PLATE 44

The Nervous System

The nervous system

The nervous system is composed of two integrated subdivisions that are responsible for conducting and processing sensory and motor information: the **central nervous system** (CNS) and the **peripheral nervous system** (PNS), which connects the CNS to the rest of the body.

The CNS includes the **brain** and **spinal cord**, which are covered by protective membranes called meninges (dura mater, arachnoid mater, and pia mater). The brain processes and coordinates all neural signals received from the spinal cord as well as its own nerves, such as the olfactory and optic nerves. It also performs complex mental functions such as thinking and learning.

The peripheral nervous system transmits input gathered from the sensory organs to the CNS. Motor output signals are relayed back to the PNS and on to the body's muscles and glands. The PNS has three separate divisions called the autonomic, sensory and motor nervous systems.

The functional units of the nervous system are **neurons**. **Sensory neurons** communicate information from sensory receptors to the CNS. **Motor neurons** relay signals from the CNS to effector (muscle and gland) cells. **Interneurons** coordinate and integrate sensory inputs and motor outputs. **Glial cells** also make up a significant portion of the nervous system and provide important support for neuron activity.

The spinal cord and nerves

The spinal cord connects the peripheral nervous system to the brain, coordinates simple reflexes to stimuli, and helps regulate the internal organs. It contains 31 pairs of **spinal nerves**, which include both sensory and motor axons. Nerve signals generated by the sensory neurons travel through the spinal cord to the brain. Signals from the motor areas of the brain are sent back through the cord and directed to the motor neurons, triggering a response.

The inner core of the spinal cord is gray matter composed of neuron cells, glial cells, and interneurons. The outer core or white matter is made up of tracts of myelinated axons responsible for transporting nerve signals. Surrounding the spinal cord are the meninges, protective bones of the vertebral column and a cushioning layer of fat and connective tissue in the epidural space.

Structure of motor neurons

Motor neurons transmit impulses to other cells, specifically muscle fibers or glands. Each neuron consists of a **central cell body** with a nucleus and numerous fiber-like extensions called dendrites that collect and relay information to the cell body for processing. Nerve signals directed from the cell body travel towards target cells via the axon, a long extension of the cell membrane. An insulating **myelin sheath** made up of lipid-like **Schwann cells** insulates the axon. Spaces between these cells are called the **nodes of Ranvier**. The **axon** branches into terminal fibers which end in **presynaptic knobs** where neurotransmitter molecules are stored.

What are synaptic connections?

Neurons in the CNS create thousands of input and output connections with other neurons, forming dense networks within the brain. Fiber-like structures called **dendrites** extend from the membrane of each neuron to receive and transmit signals from other neurons into the cell body. A long, tube-like extension called the **axon** sends signals from the neuron towards nearby target cells. As impulses arrive at the tip of the axon, the terminal bulbs release "messenger" molecules called **neurotransmitters**. These highly specialized chemicals carry nerve impulses across the tiny space between the axon and the adjacent neurons or cells, either inhibiting or activating neural impulses in the target cell.

The brain

An average brain weighs between 3 and 3.5 lbs. and is composed of over 100 billion neurons. A brain is divided into 2 structures, the larger being the **cerebrum** (80% of brain mass) and the smaller called the **cerebellum** (20% of brain mass). The cerebrum consists of 2 hemispheres, with 5 lobes in each hemisphere. Cerebral hemispheres and lobes each have the specificity of brain function. The cerebral hemispheres control the higher brain functions such as memory, speech and vision, while the cerebellum controls balance and coordination.

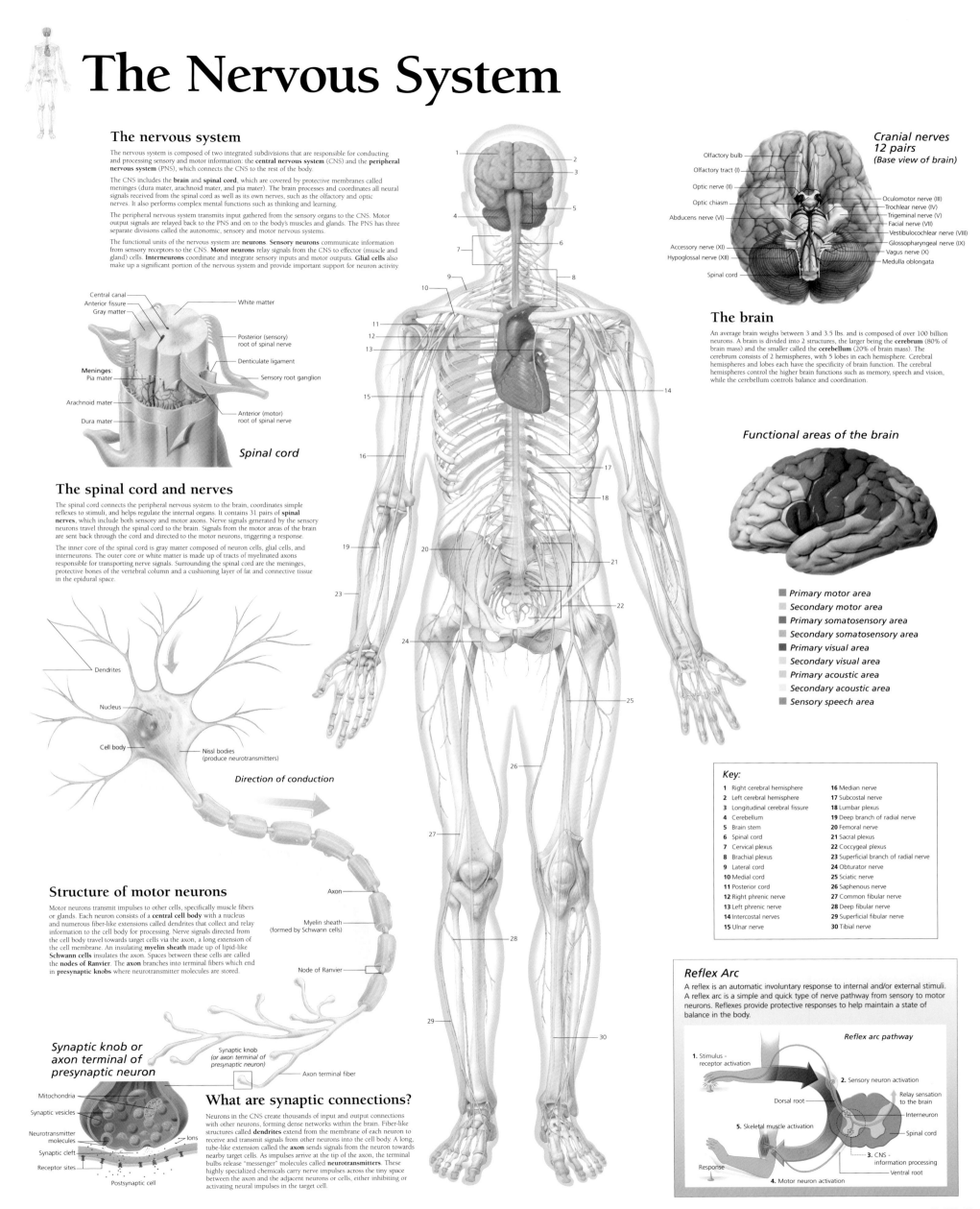

Spinal cord

Central canal
Anterior fissure
Gray matter
White matter
Posterior (sensory) root of spinal nerve
Denticulate ligament
Meninges:
Pia mater
Sensory root ganglion
Arachnoid mater
Anterior (motor) root of spinal nerve
Dura mater

Structure of motor neurons (labels)

Dendrites
Nucleus
Cell body
Nissl bodies (produce neurotransmitters)
Direction of conduction
Axon
Myelin sheath (formed by Schwann cells)
Node of Ranvier

Synaptic knob or axon terminal of presynaptic neuron

Mitochondria
Synaptic vesicles
Neurotransmitter molecules
Synaptic cleft
Receptor sites
Ions
Postsynaptic cell
Synaptic knob (or axon terminal of presynaptic neuron)
Axon terminal fiber

Cranial nerves
12 pairs
(Base view of brain)

Olfactory bulb
Olfactory tract (I)
Optic nerve (II)
Optic chiasm
Abducens nerve (VI)
Accessory nerve (XI)
Hypoglossal nerve (XII)
Spinal cord
Oculomotor nerve (III)
Trochlear nerve (IV)
Trigeminal nerve (V)
Facial nerve (VII)
Vestibulocochlear nerve (VIII)
Glossopharyngeal nerve (IX)
Vagus nerve (X)
Medulla oblongata

Functional areas of the brain

■ Primary motor area
☐ Secondary motor area
■ Primary somatosensory area
☐ Secondary somatosensory area
■ Primary visual area
☐ Secondary visual area
☐ Primary acoustic area
☐ Secondary acoustic area
■ Sensory speech area

Key:

1 Right cerebral hemisphere	**16** Median nerve
2 Left cerebral hemisphere	**17** Subcostal nerve
3 Longitudinal cerebral fissure	**18** Lumbar plexus
4 Cerebellum	**19** Deep branch of radial nerve
5 Brain stem	**20** Femoral nerve
6 Spinal cord	**21** Sacral plexus
7 Cervical plexus	**22** Coccygeal plexus
8 Brachial plexus	**23** Superficial branch of radial nerve
9 Lateral cord	**24** Obturator nerve
10 Medial cord	**25** Sciatic nerve
11 Posterior cord	**26** Saphenous nerve
12 Right phrenic nerve	**27** Common fibular nerve
13 Left phrenic nerve	**28** Deep fibular nerve
14 Intercostal nerves	**29** Superficial fibular nerve
15 Ulnar nerve	**30** Tibial nerve

Reflex Arc

A reflex is an automatic involuntary response to internal and/or external stimuli. A reflex arc is a simple and quick type of nerve pathway from sensory to motor neurons. Reflexes provide protective responses to help maintain a state of balance in the body.

Reflex arc pathway

1. Stimulus - receptor activation
2. Sensory neuron activation
Relay sensation to the brain
Dorsal root
Interneuron
5. Skeletal muscle activation
Spinal cord
3. CNS - information processing
Ventral root
Response
4. Motor neuron activation

PLATE 45

©Scientific Publishing Ltd., Elk Grove Village, IL USA
#2700

Understanding Skin

Inside the skin

The skin is a highly elastic organ covering the entire outer surface of the body. It performs numerous functions essential to survival, including prevention of **fluid loss** from body tissues; protection against **environmental toxins and microorganisms**; reception of **heat**, **cold**, and **pain sensations**; regulation of normal **body temperature**; and maintenance of **calcium levels**.

The three basic layers within the skin are the **epidermis**, **dermis** and **subcutaneous** layers.

Epidermis. The thin uppermost layer consists of basal cells, melanocytes responsible for skin color, keratin-producing cells (for hair, nails, and outer protective skin surfaces), Langerhans cells (important in immune protection) and Merkel cells (involved in sensation).

Dermis. The dense middle layer contains the skin's structural components: nerves, blood vessels, sweat glands, hair follicles, sebaceous glands and collagen.

Subcutaneous. The underlying layer of fat cells cushions body tissues from trauma, insulates against cold and stores fuel reserves.

Hair shaft

Hair shaft

- Medulla
- Cortex
- Cuticle

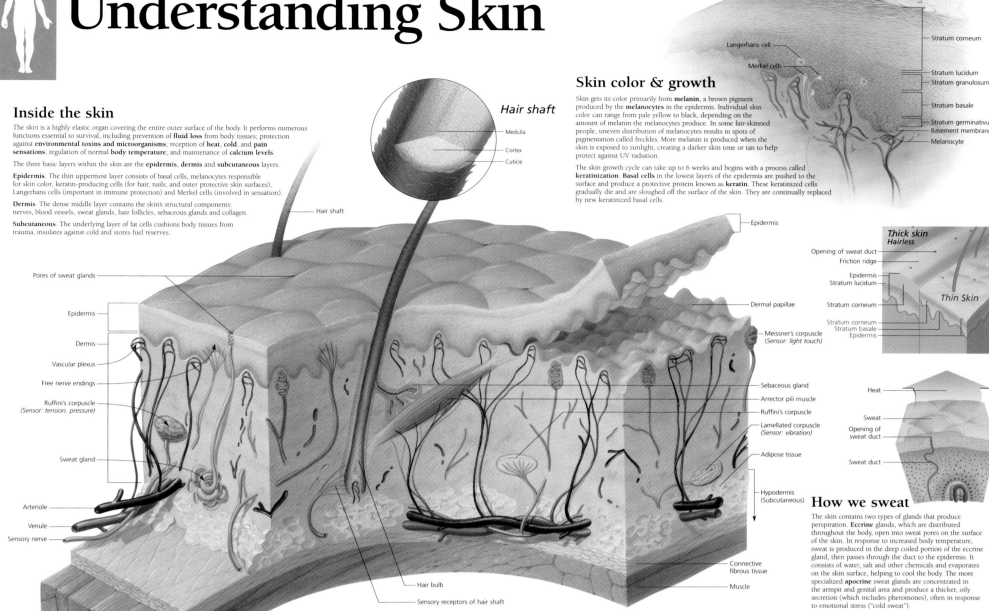

Labels: Hair shaft; Pores of sweat glands; Epidermis; Dermis; Vascular plexus; Free nerve endings; Ruffini's corpuscle *(Sensor: tension, pressure)*; Sweat gland; Arteriole; Venule; Sensory nerve; Hair bulb; Sensory receptors of hair shaft; Epidermis; Dermal papillae; Meissner's corpuscle *(Sensor: light touch)*; Sebaceous gland; Arrector pili muscle; Ruffini's corpuscle; Lamellated corpuscle *(Sensor: vibration)*; Adipose tissue; Hypodermis (Subcutaneous); Connective fibrous tissue; Muscle

Skin color & growth

Skin gets its color primarily from **melanin**, a brown pigment produced by the **melanocytes** in the epidermis. Individual skin color can range from pale yellow to black, depending on the amount of melanin the melanocytes produce. In some fair-skinned people, uneven distribution of melanocytes results in spots of pigmentation called freckles. More melanin is produced when the skin is exposed to sunlight, creating a darker skin tone or tan to help protect against UV radiation.

The skin growth cycle can take up to 6 weeks and begins with a process called **keratinization**. **Basal cells** in the lowest layers of the epidermis are pushed to the surface and produce a protective protein known as **keratin**. These keratinized cells gradually die and are sloughed off the surface of the skin. They are continually replaced by new keratinized basal cells.

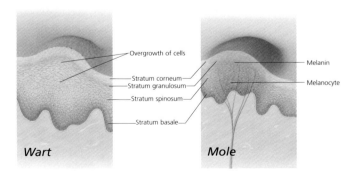

Labels: Langerhans cell; Merkel cells; Stratum corneum; Stratum lucidum; Stratum granulosum; Stratum basale; Stratum germinativum; Basement membrane; Melanocyte

Thick skin
Hairless

Labels: Opening of sweat duct; Friction ridge; Epidermis; Stratum lucidum; Stratum corneum; *Thin Skin*; Stratum corneum; Stratum basale; Epidermis

How we sweat

The skin contains two types of glands that produce perspiration. **Eccrine** glands, which are distributed throughout the body, open into sweat pores on the surface of the skin. In response to increased body temperature, sweat is produced in the deep coiled portion of the eccrine gland, then passes through the duct to the epidermis. It consists of water, salt and other chemicals and evaporates on the skin surface, helping to cool the body. The more specialized **apocrine** sweat glands are concentrated in the armpit and genital area and produce a thicker, oily secretion (which includes pheromones), often in response to emotional stress ("cold sweat").

Labels: Heat; Sweat; Opening of sweat duct; Sweat duct

Nail anatomy

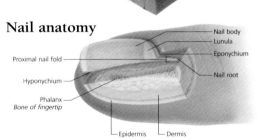

Labels: Nail body; Lunula; Eponychium; Nail root; Nail fold; Proximal nail fold; Hyponychium; Phalanx *Bone of fingertip*; Epidermis; Dermis

Like the hair, nails are an accessory structure of the skin. They contain plates of densely packed, keratinized epidermal cells which arise from superficial cells in the **nail matrix**, located under the skin behind the **nail root**. Above the nail root is the visible portion of the nail, called the **nail body**. The **free edge** of the nail extends from the nail body beyond the end of the finger or toe. Near the nail root is the **cuticle** or **lunula**, shaped like a half moon.

How wounds heal

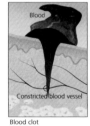 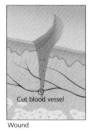  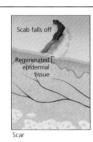

Panel labels: Blood; Scab falls off; Exudate; Regenerated epidermal tissue; Cut blood vessel; Constricted blood vessel; Granulation tissue: • Fibroblast • Lymphocyte • Macrophage; Wound; Blood clot; Scab; Scar

After a wound occurs, the damaged portion of the skin begins to heal through a series of complex overlapping stages. In the first stage, a **blood clot** forms to stop bleeding and in most cases, dries to form a protective **scab**. Below the surface, **inflammation** takes place as nearby blood vessels enlarge and deliver oxygen and nutrient-rich blood and leukocytes to cleanse the site of dead tissue and bacteria. Rapid proliferation (**regeneration**) and migration of new epithelial cells helps to replace the damaged area with new granulation tissue and close the wound. A less obvious **scar** is created when the edges of a wound are kept closed together during the healing process, allowing less granulated tissue to develop.

Burns

Burns are injuries to the skin that damage or destroy the skin's protective covering and functions. They can be caused by heat (contact with hot objects, scalding or flames), ultraviolet light (sun or artificial tanning), chemicals, electricity or even frostbite.

First degree burns are limited to the epidermis. Symptoms include heat, pain and reddening with minimal blistering or scarring.

Second degree burns may be superficial or deep but usually extend below the epidermis into the dermis, affecting the sweat glands and hair follicles. Characterized by swelling, severe pain and blistering, and red, moist skin.

Third degree burns extend from the dermis to the subcutaneous layer or underlying muscle. Charred, leathery skin may range in color from red to white or brown, with no blisters.

Burn legend:
- 1ˢᵗ degree burns — Epidermis involvement
- 2ⁿᵈ degree burns — Epidermis & dermis involvement
- 3ʳᵈ degree burns — Entire skin is destroyed

Skin and acne

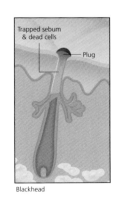

Panel labels: Hair shaft; Trapped sebum & dead cells; Infection & inflammation; Plug; Sebum; Sebaceous gland; Follicle; Sebum, dead cells & bacterial infection; Normal skin; Blackhead; Pustule

Acne vulgaris is a common condition caused by inflammation in the **sebaceous glands** and **hair follicles**. Blemishes including **comedones** (blackheads and whiteheads), **pustules** (pimples), and **nodules** or **cysts** can appear on the face, neck or upper body. Usually beginning in adolescence, acne can last from 5 to 10 years. In some cases, acne may continue into adulthood or occur for the first time in adults.

How acne develops:

- **Hormones** trigger an increase in the production of **sebum** (an oily substance important to skin lubrication) in the sebaceous glands.
- Excess sebum and dead skin cells become trapped in tiny **hair follicles** located near or within the sebaceous glands.
- **Bacteria** from the skin enter the clogged follicles and multiply, causing inflammation, pus and swelling (visible as comedones or pimples).
- In some cases, pustules may progress to nodules or cysts that extend deeper into the skin and cause scarring.

Warts & moles

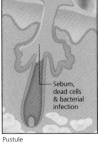 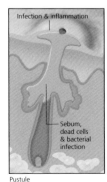

Labels: Overgrowth of cells; Stratum corneum; Stratum granulosum; Stratum spinosum; Stratum basale; Melanin; Melanocyte; *Wart*; *Mole*

Warts are hard, benign lumps on the surface of the skin, usually with a rough, raised surface and round or oval growth. Warts are produced when a **virus** enters the topmost layer of skin, causing an overgrowth of skin cells. Most common in children, they can be spread through **direct skin contact** and typically appear on the face, hands, or feet (often as **plantar** warts). Warts usually cause no discomfort and disappear within two years. However, treatments including medications, cryotherapy or electrocautery may be used to remove warts more quickly.

Moles are skin lesions common in light-skinned people that are often small and round and usually benign. Moles contain **melanin**, which gives them a brown or tan color. Also called **nevi** (singular: nevus), moles can range in size from tiny to very large and may have smooth or irregular borders. Unusual changes in the size or appearance of a nevus can be an important warning sign of **melanoma** (skin cancer).

Skin cancer

Skin cancer is a malignant growth on the skin caused by the uncontrolled growth of epidermal cells. Skin cancer is associated with known **risk factors** including **sun exposure** and sunburn, family **history of skin cancer**, **light** or **pale complexion**, and age. Lesions may be small, shiny, waxy, crusty, or rough, asymmetrical in texture, or have an irregular border. Cancers are often larger than 6mm in size and can range in color from white to blue, brown, or black. All unusual or suspicious skin lesions should be examined by a physician.

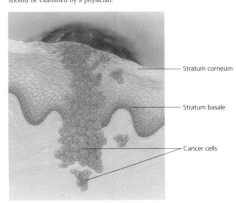

Labels: Stratum corneum; Stratum basale; Cancer cells

©Scientific Publishing Ltd., Elk Grove Village, IL USA
#2500

PLATE 46

The Male Reproductive System

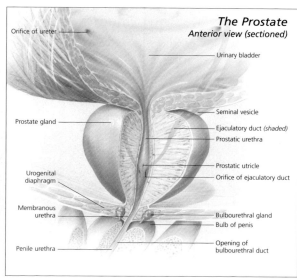

The Prostate
Anterior view (sectioned)

Orifice of ureter
Urinary bladder
Prostate gland
Seminal vesicle
Ejaculatory duct *(shaded)*
Prostatic urethra
Prostatic utricle
Urogenital diaphragm
Orifice of ejaculatory duct
Membranous urethra
Bulbourethral gland
Bulb of penis
Penile urethra
Opening of bulbourethral duct

What is the prostate?

The prostate is a small gland located beneath the bladder and just in front of the rectum, behind the base of the penis. The prostate is similar to a walnut in both shape and size and surrounds the upper portion of the **urethra**, which passes through it. The primary purpose of the prostate is the production of fluid for semen. It also functions as a valve, preventing the leakage of urine from the bladder and the entry of sperm and seminal fluid into the bladder.

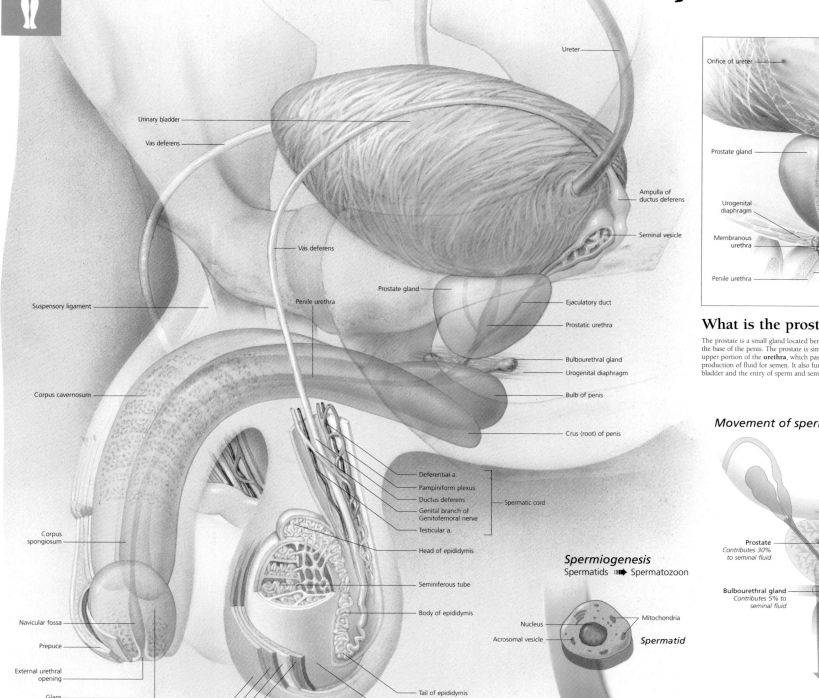

Ureter
Urinary bladder
Vas deferens
Ampulla of ductus deferens
Seminal vesicle
Vas deferens
Prostate gland
Ejaculatory duct
Prostatic urethra
Suspensory ligament
Penile urethra
Bulbourethral gland
Urogenital diaphragm
Corpus cavernosum
Bulb of penis
Crus (root) of penis
Deferential a.
Pampiniform plexus
Ductus deferens
Genital branch of Genitofemoral nerve
Testicular a.
Spermatic cord
Head of epididymis
Corpus spongiosum
Seminiferous tube
Body of epididymis
Navicular fossa
Prepuce
External urethral opening
Glans
Skin
Dartos muscle
Superficial scrotal fascia
Cremaster muscle
Tail of epididymis
Testis (covered by visceral layer of tunica vaginalis)

Male reproductive system

The adult male reproductive system is comprised of two primary external structures, the **testes** and the **penis**. The testes are a pair of organs approximately 1.5 inches in length and 1 inch in diameter. They are each divided into hundreds of tiny compartments or **lobules** containing the **seminiferous tubules**, tightly coiled structures where **spermatogenesis** and **spermiogenesis** take place.

The testes are encased in a protective saclike structure called the **scrotum**, which provides support and regulates the position of the testes relative to the body. The scrotum is divided into two compartments that separate the testes from each other, preventing injury or infection on one side from affecting the other. The inner muscle layers of the scrotum react to changes in external temperature to maintain the proper temperature within the testes for spermatogenesis. The testes are relaxed and lowered during warm temperatures and contracted and elevated when cold.

Testosterone is produced within the testes by specialized cells known as **interstitial cells**. Testosterone is the male hormone responsible for maintaining the structure and function of the sex organs and promoting the development of male secondary sex characteristics.

Movement of spermatozoa

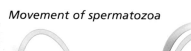

Seminal vesicle
Contributes 60% to seminal fluid
Prostate
Contributes 30% to seminal fluid
Ejaculatory duct
Ductus deferens
Bulbourethral gland
Contributes 5% to seminal fluid
Penile urethra
Ejaculate
Spermatozoa movement by peristaltic action of the ductus deferens
Head of epididymis
Epididymis and sustentacular cells
Contribute 5% to seminal fluid
Tail of epididymis

Spermiogenesis
Spermatids ➡ Spermatozoon

Nucleus
Acrosomal vesicle
Mitochondria
Spermatid
Acrosomal vesicle
Flagellum

Accessory glands
• Seminal vesicles
• Prostate
• Bulbourethral glands

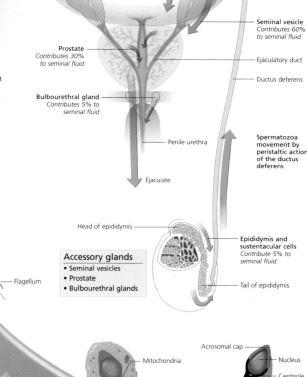

Acrosomal cap
Shed cytoplasm
Mitochondria
Residual cytoplasm
Flagellum
Acrosomal cap
Nucleus
Centriole
Mitochondrial spiral
Fibrous sheath
Spermatozoon

Spermiogenesis

During **spermatogenesis**, sperm cells are produced in the seminiferous tubules. Sperm cells first differentiate into primary spermatocytes before undergoing meiosis and producing a pair of secondary spermatocytes. As each secondary spermatocyte divides, a pair of spermatids is produced. In the next phase, **spermiogenesis**, the spermatids are embedded within large cells called **sustentacular cells**, where they continue to mature. Spermatids undergo a dramatic change in form during spermiogenesis, gradually developing the structure and appearance of functional **spermatozoa**. Newly produced spermatozoa detach from the sustentacular cells and are transported to the **epididymis**, a coiled thin-walled tube within the scrotum that connects the testis to the **vas deferens** (the duct through which sperm travel to the prostate before ejaculation).

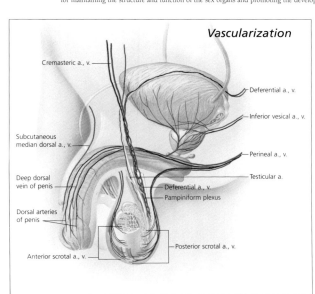

Vascularization

Cremasteric a., v.
Deferential a., v.
Inferior vesical a., v.
Subcutaneous median dorsal a., v.
Perineal a., v.
Deep dorsal vein of penis
Testicular a.
Deferential a., v.
Pampiniform plexus
Dorsal arteries of penis
Posterior scrotal a., v.
Anterior scrotal a., v.

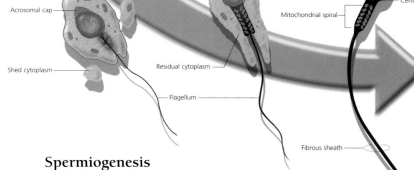

Urinary bladder
Vas deferens
Cut and suture
Cut and suture
Epididymis
Testis

Vasectomy

Vasectomy is a surgical procedure that involves cutting and sealing the vas deferens, or ducts that transport sperm from the testes. Small incisions are made in the skin of the scrotum and approximately one half-inch is removed from the vas deferens in each testis. The ends are sutured or cauterized and placed back in the scrotum. It is possible to reverse the procedure by reattaching the vas deferens, but more extensive surgery is required and only 30 to 40 percent of men who undergo vasectomy reversal may be able to successfully father children.

©Scientific Publishing Ltd., Elk Grove Village, IL., USA
#4000

PLATE 47

Understanding the Prostate

The prostate
Anterior view (sectioned)

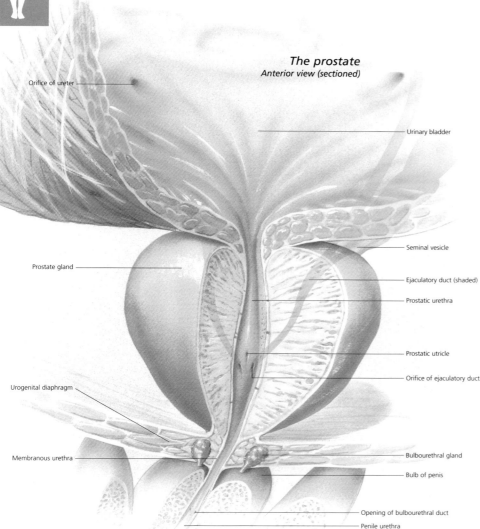

Orifice of ureter
Prostate gland
Urogenital diaphragm
Membranous urethra
Urinary bladder
Seminal vesicle
Ejaculatory duct (shaded)
Prostatic urethra
Prostatic utricle
Orifice of ejaculatory duct
Bulbourethral gland
Bulb of penis
Opening of bulbourethral duct
Penile urethra

What is the prostate?

The prostate is a small gland located beneath the bladder and just in front of the rectum, behind the base of the penis. The prostate is similar to a walnut in both shape and size and surrounds the upper portion of the **urethra**, which passes through it. The primary purpose of the prostate is the production of fluid for semen. It also functions as a valve, preventing the leakage of urine from the bladder and the entry of sperm and seminal fluid into the bladder.

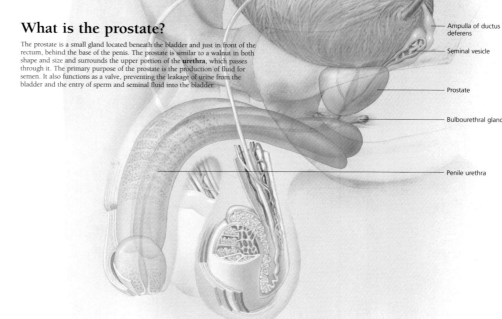

Ureter
Ductus deferens
Bladder
Ampulla of ductus deferens
Seminal vesicle
Prostate
Bulbourethral gland
Penile urethra

Prostate — zones
Sagittal section

Zones of the prostate

There are three primary zones within the prostate. The outermost section is called the **peripheral zone**. It makes up approximately 70% of the prostate gland's total volume and is the area where prostate cancer is most likely to develop. The innermost section of the prostate is the **transition zone**, a small area that surrounds the urethra. In **benign prostatic hyperplasia** or BPH (see below), noncancerous growth of tissue (hyperplasia) in the transition zone causes constriction of the urethra and restricts urinary flow. Between these two zones is the **central zone** of the prostate, through which the ejaculatory ducts pass.

Zone key:
- Central zone
- Peripheral zone
- Transition zone
- Anterior zone

What is prostatitis?

Prostatitis is a painful condition involving infection or inflammation of the prostate gland. The most common of all prostate diseases, its symptoms include pain between the rectum and testicles, in the groin and genital area, and in the lower back. There are several types of prostatitis, including **acute** and **chronic**, **bacterial** and **nonbacterial**. It is typically diagnosed by urinalysis and treated with antibiotics, anti-inflammatory drugs, and other medications. Unfortunately, the causes of chronic prostatitis, the most common form of the disease, are not yet understood. Research into improved treatments for prostatitis is ongoing.

Bladder
Seminal vesicle
Inflamed prostatic tissue
Urethra
Ejaculatory duct
Prostate
Bulbourethral gland
Urogenital diaphragm
Sagittal section

Detection and treatment of prostate cancer

Recent improvements in detection methods are now allowing earlier diagnosis and treatment of prostate cancer. This has produced a significant decline in mortality rates in recent years.

Bladder
Prostatic urethra
Prostate
Cancerous tumor
Sagittal section

Symptoms

Although early stages of prostate cancer often go unnoticed, a variety of symptoms may occur as the disease progresses. Many of these symptoms are also associated with benign prostatic hyperplasia (BPH), a noncancerous condition.

- Frequent urination, particularly at night
- Inability to urinate or difficulty starting urination
- Pain or burning during urination
- Presence of blood in urine or semen
- Difficulty achieving an erection
- Pain during ejaculation
- Chronic pain or stiffness in the lower back or legs

What is benign prostatic hyperplasia (BPH)?

BPH, also known as **enlarged prostate**, is a noncancerous growth of tissue within the transition zone that can restrict urination and cause other urinary problems. It is a common condition in men over 60 years of age. BPH is **not associated with prostate cancer** and can be effectively treated. About one third of all men with BPH will eventually require treatment for their symptoms.

BPH symptoms and treatment

BPH symptoms include a **weak urine stream**, a sensation of **incomplete emptying** of the bladder, **urinary frequency** and **urgency**, and the need to **urinate several times during the night**.

Diagnosis of BPH is based on factors including medical history, a physical exam, and a symptom score assessment. **Blood tests**, **urinalysis**, **X-rays**, **cystoscopy**, **ultrasound** and other tests may be used to confirm the diagnosis and guide treatment decisions.

Treatment options include:
- **Watchful waiting** — if symptoms are manageable
- **Medication therapy** — to relax the prostate muscles, shrink prostatic tissue, or both
- **Minimally invasive therapies** — such as microwave thermotherapy heat targeted areas of prostate tissue to relieve symptoms
- **TURP** — or transurethral resection of the prostate involves the surgical removal of the inner portion of the prostate
- **Phytotherapeutics** — saw palmetto and other plant-based therapies may be effective in treating BPH but are not FDA approved

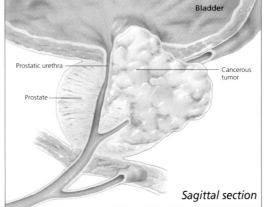

Bladder
Prostate
Prostatic urethra
Noncancerous tumor
Sagittal section

Detection

The most reliable detection of early prostate cancer involves a combination of two tests:
- **Digital rectal exam** — a physical examination of the prostate via the rectum. It detects hardness, bumps, or swelling caused by cancer or other prostate problems. While this test is important, it often cannot detect prostate cancer until it is more advanced.
- **Prostate specific antigen (PSA)** — a blood test that measures an enzyme produced by the prostate. An elevated PSA is considered the best predictor of prostate cancer and can detect the presence of disease up to 6 years earlier than a digital rectal exam.

If either test indicates a possibility of cancer, additional tests may be performed to confirm the diagnosis:
- **Transrectal ultrasonography** — to measure prostate size and locate possible sites of cancer cells for needle biopsy.
- **Needle biopsy** — tissue samples are taken from several areas of the prostate for microscopic diagnosis.

Digital rectal exam

Treatment

The type of treatment recommended varies from patient to patient. A classification method called the **Gleason score** is often used to rank malignancy and determine the most appropriate treatment.
- **Watchful waiting** — recommended for some men with early, slow-growing prostate cancer or other serious medical problems.
- **Surgery** — options for more advanced disease include radical prostatectomy (removal of the entire prostate gland) and nerve-sparing surgery, a newer surgical technique.
- **Radiation** — external beam radiation therapy offers a highly effective alternative to surgery in many prostate cancer patients.
- **Brachytherapy** — radioactive seeds are implanted in the prostate to eradicate cancer cells.
- **Hormone therapy** — generally recommended when prostate cancer has spread to other tissues. Hormones linked to the growth of cancer cells are inhibited through surgical or drug therapy.

©Scientific Publishing Ltd., Elk Grove Village, IL USA
#4500

PLATE 48

The Female Reproductive System

Vagina, uterus, uterine tubes, ovaries
(Coronal section)

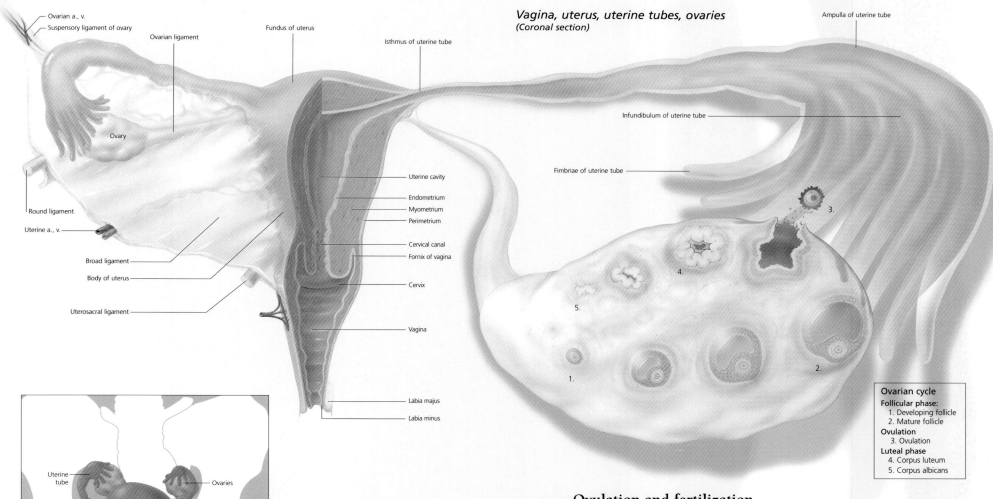

Ovarian a., v.
Suspensory ligament of ovary
Ovarian ligament
Fundus of uterus
Isthmus of uterine tube
Ampulla of uterine tube
Infundibulum of uterine tube
Ovary
Fimbriae of uterine tube
Round ligament
Uterine a., v.
Broad ligament
Body of uterus
Uterosacral ligament
Uterine cavity
Endometrium
Myometrium
Perimetrium
Cervical canal
Fornix of vagina
Cervix
Vagina
Labia majus
Labia minus

1. 2. 3. 4. 5.

Ovarian cycle
Follicular phase:
1. Developing follicle
2. Mature follicle
Ovulation
3. Ovulation
Luteal phase
4. Corpus luteum
5. Corpus albicans

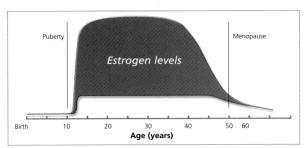

Uterine tube
Ovaries
Uterus
Bladder
Vagina

Female reproductive system

The internal structures of the adult female reproductive system include the vagina, cervix, uterus, ovaries and uterine tubes. These structures form a pathway allowing the release of **ova** (egg cells), fertilization by **sperm** and delivery of a developed fetus.

The **vagina** is a muscular tube approximately 3 to 4 inches long. It extends from the external genital organs to the **cervix**, the lower part of the uterus. A channel through the cervix allows the passage of sperm and menstrual discharge.

The **uterus**, a hollow pear-shaped organ with muscular walls that contract during childbirth, consists of three sections: the cervix, a tapered middle section known as the **corpus** (main body) and the wide **fundus** at the top of the uterus. The inner lining of the corpus is the **endometrium**, which thickens during each menstrual cycle in preparation for implantation by a fertilized egg. This lining is shed during menstruation if fertilization does not occur.

The **ovaries** are located on each side of the uterus. These small oval glands are similar to an almond in size and shape. They are responsible for producing ova as well as the female sex hormones **estrogen** and **progesterone**. The ovaries are connected to the uterus by the fallopian tubes, narrow ducts approximately 2 to 3 inches in length that provide a passageway for the ova to reach the uterus. Fertilization normally occurs in the uterine tubes.

Ovulation and fertilization

Fertilization and activation of oocyte

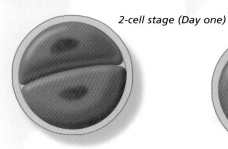

2-cell stage (Day one)

4-cell stage (Day two)

Ovulation is the release of a single mature ovum from one of the ovaries, which is triggered by a sudden rise in the blood level of the gonadotrophic hormone LH. The ovum travels down the uterine tube and enters the uterine cavity. An unfertilized ovum passes out of the body through the vagina. Ovulation normally occurs around day 14 of the cycle.

Fertilization is the fusion of genetic material from a mature ovum with that from a mature sperm to produce a fertilized egg (or zygote). Fertilization normally occurs in the uterine tube. The zygote begins to divide as it travels down the uterine tube towards the uterus, eventually forming an embryo that may successfully implant into the endometrium.

Nucleus
Polar body
Zona pellucida
Corona radiata

Oocyte at ovulation

What is estrogen?

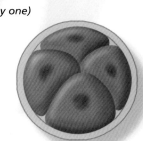

Puberty
Menopause
Estrogen levels
Birth   10   20   30   40   50   60
Age (years)

Estrogen is the female sex hormone. There are actually several different estrogens, the most important being **estradiol**. Estrogens are produced in the **ovaries**, which are located in the lateral walls of the pelvis on either side of the **uterus** (womb). The two ovaries produce female sex cells (the ova or 'eggs') and hormones (estrogens). The ovaries do not begin to produce estrogens until the onset of puberty. Estrogens are responsible for the appearance of female **secondary sexual characteristics**, which enable a young woman to achieve full reproductive fertility (e.g. growth of the reproductive organs, onset of menstrual periods, etc.).

Estrogens affect the female reproductive organs and other parts of the body –

- **Genital tract:** estrogens stimulate a favorable environment for the survival of sperm during the **menstrual cycle**
- **Breast tissue:** estrogens stimulate the growth of non-glandular breast tissue
- **Heart:** estrogens improve circulation and prevent high blood pressure
- **Skeleton:** estrogens help retain calcium in the bones

What is the menstrual cycle?

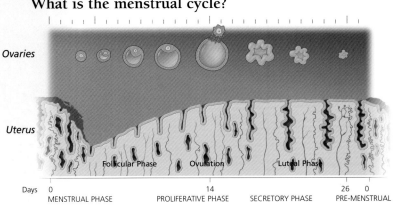

Ovaries
Uterus

Follicular Phase   Ovulation   Luteal Phase

Days   0   14   26   0
MENSTRUAL PHASE   PROLIFERATIVE PHASE   SECRETORY PHASE   PRE-MENSTRUAL

The **menstrual cycle** refers to the sequence of events that occurs in the cell layers lining the uterus (womb), called the **endometrial layer** or **endometrium**; it normally lasts for approximately 28 days. The purpose of the menstrual cycle is to prepare the uterus for possible pregnancy. It is intimately linked to events occurring in the ovary that prepare a mature ovum (or egg cell) for release (ovulation) and possible fertilization; this is known as the ovarian cycle. Hormones produced in the ovaries, namely **estrogens** and **progesterone**, control the menstrual cycle. Hormones produced in the brain control the ovarian cycle; these hormones are called **gonadotrophins** (LH and FSH).

During menstruation the cells lining the uterus (endometrium) become detached from the uterine wall and fall off. This is accompanied by bleeding and normally lasts around 5 days. The tissue pieces and blood pass into the vagina and out of the body. Menstruation marks the **beginning** of the menstrual cycle.

©Scientific Publishing Ltd., Elk Grove Village, IL. USA
#5000

PLATE 49

Understanding Menopause

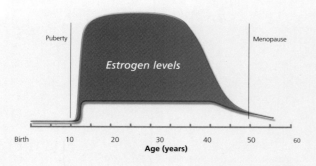

Estrogen levels

Puberty — Menopause

Birth | 10 | 20 | 30 | 40 | 50 | 60
Age (years)

The role of estrogen

As menopause approaches, levels of the female sex hormone **estrogen** gradually decline. There are several different estrogens, including the most potent form, **estradiol**. Estrogens and female sex cells (the *ova* or **eggs**) are produced in the ovaries from the onset of puberty through menopause. Estrogens are responsible for the development and maintenance of **female characteristics** and **sexual reproduction**. They also play important roles in other parts of the body:

- **Genital tract**   Stimulate a favorable environment for the survival of sperm during the menstrual cycle
- **Breasts**   Stimulate the growth of non-glandular breast tissue
- **Heart**   Improve circulation and prevent high blood pressure
- **Skeleton**   Help retain calcium in the bones

What is menopause?

It is a gradual process that occurs as a woman's ovaries produce decreasing amounts of estrogen until menstruation and ovulation finally cease. It normally occurs between the ages of approximately 40 and 55. Menopause may also result from surgical removal of the ovaries.

Menopause can be divided into three stages:

Perimenopause – A period of several years during which estrogen and progesterone levels steadily decline. Symptoms may include changes in the menstrual cycle (length and flow); intermittent ovulation; missed periods; hot flashes; and emotional/cognitive changes.

Menopause – Menopause occurs when ovarian hormone production is too low to initiate a menstrual cycle and periods stop completely.

Postmenopause – Also known as true menopause, this phase is usually defined as 12 full months without a menstrual period. All estrogen production in the ovaries has ceased.

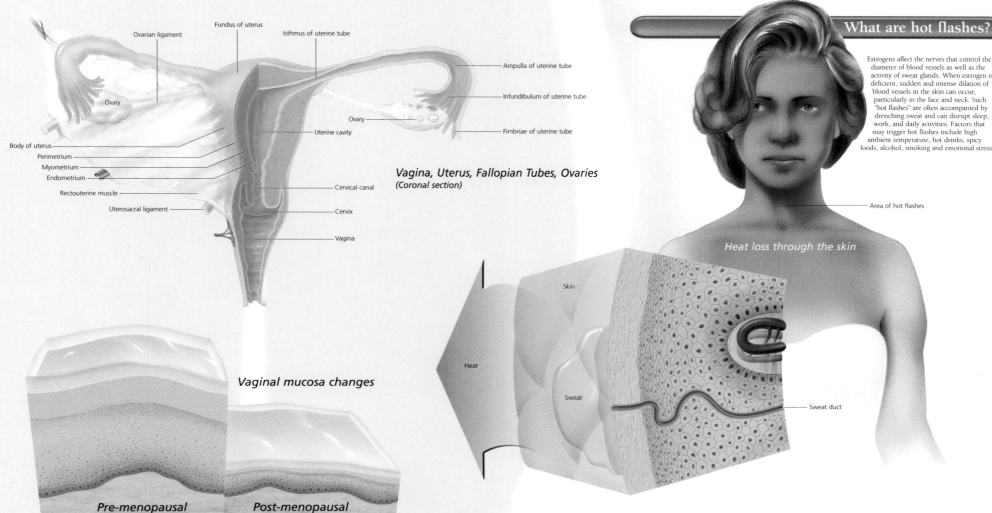

Ovarian ligament · Fundus of uterus · Isthmus of uterine tube · Ampulla of uterine tube · Infundibulum of uterine tube · Ovary · Ovary · Fimbriae of uterine tube · Uterine cavity · Body of uterus · Perimetrium · Myometrium · Endometrium · Cervical canal · Rectouterine muscle · Cervix · Uterosacral ligament · Vagina

Vagina, Uterus, Fallopian Tubes, Ovaries
(Coronal section)

What are hot flashes?

Estrogens affect the nerves that control the diameter of blood vessels as well as the activity of sweat glands. When estrogen is deficient, sudden and intense dilation of blood vessels in the skin can occur, particularly in the face and neck. Such "hot flashes" are often accompanied by drenching sweat and can disrupt sleep, work, and daily activities. Factors that may trigger hot flashes include high ambient temperature, hot drinks, spicy foods, alcohol, smoking and emotional stress.

Area of hot flashes

Heat loss through the skin

Skin · Heat · Sweat · Sweat duct

Vaginal mucosa changes

Pre-menopausal   *Post-menopausal*

How does the body change during menopause?

One of the predominant changes that occurs with menopause is the gradual thinning and drying (*atrophy*) of tissue in the genitourinary tract, including the vagina, vulva and urethra. Other physiological changes triggered by decreased estrogen include:

- **Reproductive system**   Organs decrease in size; production of eggs and hormones stops
- **Breasts**   Loss of firmness; change in size/shape
- **Pelvis**   Supporting ligaments may weaken and cause urinary incontinence
- **Skin & hair**   Often become thinner and drier; skin may lose some natural elasticity and be more prone to UV damage
- **Musculoskeletal**   Loss of bone density, which can lead to osteoporosis
- **Blood cholesterol**   Levels of LDL cholesterol rise as estrogen levels decline

Emotional and cognitive effects of menopause

Changing hormonal levels can trigger changes in mood and cognitive function. These symptoms can occur briefly or for longer durations, sometimes over several years.

- **Anxiety or panic attacks**
- **Mood swings and irritability**
- **Short-term memory problems**
- **Depression**
- **Difficulty concentrating**
- **Decreased sex drive**

Understanding the menstrual cycle

Fertilization
The fusion of genetic material from a mature ovum and a mature sperm produces a fertilized egg (or *zygote*). Fertilization normally occurs in the uterine tube. As the zygote begins to divide, it travels down the uterine tube towards the uterus, eventually forming an **embryo** that may successfully implant into the lining of the uterus (endometrium).

Menstruation
During menstruation, cells lining the uterus (*endometrium*) become detached from the uterine wall and fall off. Blood and tissue from the uterine lining are discharged through the vagina, a process which normally lasts around 5 days. Menstruation marks the **beginning** of the menstrual cycle.

Ovulation
Approximately **14 days** into the menstrual cycle, a sudden rise in the blood level of gonadatrophic hormone LH triggers the release of a single mature *ovum* or egg from one of the ovaries. The egg travels down the uterine tube until it reaches the uterine cavity. An unfertilized egg is discharged through the vagina.

Treating menopause symptoms and risks

There are a variety of treatments available today for reducing menopause symptoms and addressing potential risks such as osteoporosis, higher cholesterol and heart disease. Decisions are highly individualized and should be made only after discussion with a physician.

HRT – Menopausal symptoms often require no treatment. If side effects are severe, HRT (**hormone replacement therapy**) may be recommended to reduce the intensity and incidence of hot flashes, alleviate insomnia, and decrease vaginal dryness and drying/thinning of the skin. HRT may also decrease the risks of postmenopausal heart disease and osteoporosis.

Non-prescription alternatives – Other approaches to both the symptoms and longer-term effects of menopause can include **increased exercise**, **stress reduction** programs, and a nutritionally balanced diet that includes foods rich in **calcium** and phytoestrogens, such as **soy**.

New therapies – New therapies such as **SERMs** (selective estrogen receptor modulators) are also available for the treatment of menopausal symptoms. SERMs inhibit the effects of estrogen on certain tissues while mimicking beneficial estrogen action in other parts of the body, including bone tissue and blood cholesterol.

The menstrual cycle refers to the sequence of events that occurs in the **endometrium**, the cell layers lining the uterus. This cycle normally lasts for approximately 28 days. Its purpose is to prepare the uterus for possible pregnancy. The menstrual cycle is governed by the **endocrine system**.

During the first two weeks, the pituitary gland releases follicle stimulating hormone (**FSH**) to stimulate egg growth in the ovary. Ripening eggs produce estrogen, causing thickening of the uterine lining. About 14 days into the cycle, levels of luteinizing hormone (**LH**) increase, triggering the release of a ripened egg (follicle). If fertilization does not occur, decreasing estrogen and progesterone levels initiate menstruation and the cycle begins again.

©Scientific Publishing Ltd., Elk Grove Village, IL USA
#5500

PLATE 50